*Aqua*culture *Tech*nology

CATALOG

www.aqua-tech.eu

INTRODUCTION

Imprint

3rd edition
All rights reserved
© 2018 by:

AquaTech

Unterbrunnweg 3
6370 Kitzbühel
Austria/Europe

Phone: +43-5356-71399
Fax: +43-5356-64870
Cell: +43-664-1048297

Email: aquatech@a1.net
Email: info@aquaculture-com.net
Email: fishfarming@technologist.com

http://www.aqua-tech.eu
http://www.aquaculture-com.net
http://members.a1.net/aquatech

Printed in Germany by:

Books on Demand GmbH

ISBN 978-3-902855-24-4

Distribution price € 5.-

For more information and prices, please visit our extensive web site.

Quality for reasonable prices – Our success and your profit!

INTRODUCTION

Index page

HOLDING		5
	Filters	
	Heaters	
	Basins	
BREEDING		13
	Incubators	
	Troughs	
	Jars	
FEEDING		19
	Spreaders	
	Pendulums	
	Automates	
AERATING		23
	Injectors	
	Diffusers	
	Blowers	
MONITORING		32
	Meters	
	Testers	
	Analysers	
BOAT-DOCKS		39
	Pontoons	
	Boats	
	Motors	
NET-CAGES		43
	Cages	
	Sheets	
	Nets	
POWER-PLANTS		50
	Turbines	
	Generators	
	Aggregates	
TRANSPORTING		54
	Tanks	
	Graders	
	Lifts	
PROCESSING		65
	Filleters	
	Skinners	
	Smokers	
OTHERS		76
	Fishes	
	Feeds	
	Service	

INTRODUCTION

An approved range of products for aquaculture

The need and demand for fish and seafood are rising continually worldwide, and wild fish stocks are steadily becoming scarcer and scarcer. Aquaculture is a young and innovative part of the economy with high growth rates and secured prospects.
The future of fish production requires that water be used as economically as possible, and that waste products be removed safely with the lowest possible energy consumption. Environmental legislation is progressively reducing the possibilities of open water fish production and existing fish farms are confronted with directives for purifying their waste discharges. This has accelerated the development of intensive, water-saving systems for fish production, which do not depend upon natural environmental factors. Production in closed water systems offers a more compact growing environment for fish, lower shared personnel costs and the potential for good profits. In such systems waste water that is loaded with metabolic by-products is recycled through biological- and physical purification units, and re-used.
Through improvements to these techniques we are able to offer very efficient and cost effective systems for both fresh- and seawater. Risk-minimization, process-control and fish health have all been considered in our designs, which are typically constructed in compact, modular units. Depending on the system's size and customer specifications our systems will be supplied with modern monitoring-, regulatory- and security units, and working accessories. Because of our extensive work and many years' experience in fish farming we know how important it is to have secure, stable, safe and effective units and equipment. This knowledge is apparent in our systems, machines, instruments and accessories. Beside the development and construction of new devices and units, our systems have been tested for several years before they are approved. Some of the systems are produced by us - other components are supplied by cooperating business partners.

We offer a complete program for aquaculture from the supply of single components like: incubation-, holding-, feeding-, aerating-, and processing-systems, to complete hatcheries and grow-out systems suitable for many species. If preferred, we can offer complete, all-inclusive services – from consulting and planning to training and marketing, etc. In the following pages you will find a selection of our approved range of products. Now also you can use our extensive experience and know-how to secure your success!

Please let us know if you have any special questions or requests.
I will be pleased to hear from you,
Sincerely

Fish Master
Martin Hochleithner

HOLDING

HOLDING

Recirculation systems for simple rearing

These modules are available for a yearly production-capacity of 1-2 t/year (125-250 kg standing crop) and allow our customers to become familiar with fish farming techniques without high risks. They consist of 2 holding tanks and 1 filter unit with all the necessary equipment.

Suspended solid waste in water coming from the fish tanks is removed in a swirl separator and thereafter by filter brushes prior to bio-filtration. The bacterial films needed for nitrification will develop naturally within the biofilter and mud will be carried away by the current. Our bio-reactors have a large specific surface area and therefore need only very small filter units. By introducing fine air bubbles in a special aeration- and air-lift unit, our systems operate without pumps, and therefore require less energy and reduced operating costs. A UV-sterilizer can be included. Because the units are compact, our systems can be installed anytime and anywhere without additional expense of rebuilding or extending existing buildings. Floating aquaponic-kits or standing hydroponic-sets are optionally available.

Recirculation systems for intensive production

These modules have a production-capacity of 10-50 t/year (2-10 t standing crop) and up, and can be designed to specific customer requirements.

Water from the fish tanks passes through a drum filter to remove suspended solids prior to bio-filtration, where nitrification and aeration take place. From there the water is pumped to the fish tanks, which can be selected to meet the customer's requirements in shape, size and number. This system is suitable for all fish species. For use with seawater the drum filter is exchanged for a protein skimmer, which is more effective, and also allows for the economical use of ozone.

Depending on production capacity and customer preferences, the systems can be supplied with modern control and safeguarding technique, as well as working equipment and accessories.

HOLDING

Holding aquaria for fish and seafood

An aquarium should be seen as a complex grouping of elements functioning synergetic to maintain constancy of environmental characteristics, countering any variation whatsoever and simulating as far as possible the whole biological cycle that is present in nature and indispensable to the live of the aquatic animals.

This professional range of aquariums are designed for holding of live crustaceans or fishes at a working temperature of 10-30 °C and consists of double glass walls in which the two layers of glass are separated by a dry air gap. The aquaria are delivered as complete assembled units with ALU-frame and PVC-panels (70 cm high, color: black) with cover and lights (IP67), external mechanical/biological filter, circulation pump and chiller with thermostat, as well as cable and plough.

Type	AT-70	AT-100	AT-150	AT-200
Dimensions (LxDxH)	70x50x140 cm	100x50x140 cm	150x60x140 cm	200x60x140 cm
Aquaria volume	180 l	220 l	450 l	600 l
Holding capacity	5 kg	15 kg	25 kg	35 kg

The two larger systems are also available with separate two-compartment aquaria. Single wall glass aquaria (200-600 l volume) with stainless steel frame, on request.

Holding tanks for mollusks and crustaceans

This isolated (GRP) glassfiber reinforced polyester tanks are especially suitable for the holding of Crustacea and Mollusca (like: lobsters and clams etc.).

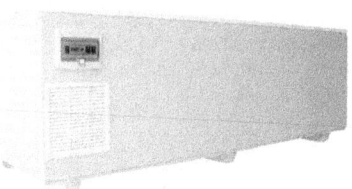

The units are delivered complete with a bottom integrated mechanical-biological filter (*Lithothamnium calcareum*) with pre-filter, titanium-chiller (R134) with digital display and thermostat, UV-C and recirculation system with PVC-pipes. They are built for an electric connection of 380 V/50 Hz and a working temperature of 5-25 °C.

Larger stacked systems, on request.

Type/Volume	A660	B990	C1320
Outside dimensions (LxWxD)	220x130x102 cm	320x130x102 cm	420x130x102 cm
Inside dimensions (LxWxD)	200x110x34 cm	300x110x34 cm	400x110x34 cm
Holding capacity (approx.)	70-140 kg	110-220 kg	150-300 kg

HOLDING

Drum filters with high capacity

These drum filters (0.4-2.0 m Ø) for the mechanical cleaning of moderately- to heavily loaded aquaculture or industrial waste water are made from stainless steel and are supplied complete with filter frames and screens, backwash pump, electronic control unit, and with or without an installation tank. They are available with filter screens ranging from 20 to 100 μm and for flow rates of 5 to 1,000 l/s. The water requirement for backwashing is 0.1 to 1.5 l/s at 3-4 bar, depending on unit size.

The following standard types are available:

Type	2-60	2-80	4-80	6-120	9-120	12-160	16-160	20-160	24-160	28-160	32-160
Diameter cm	60	80	80	120	120	160	160	160	160	160	160
Filter plates	2	2	4	6	9	12	16	20	24	28	32
Filter area m2	0.52	0.86	1.72	2.58	3.87	5.16	6.88	8.60	10.32	12.04	13.76
Motor Watts	250	250	250	370	370	550	550	550	550	550	550

Protein skimmers for aquaculture and aquaristic

The reaction- and foam tubes of these skimmers are well proportioned and have adequate volume. A special injection unit generates a vacuum, where air is drawn in and distributed as fine bubbles. In seawater each bubble has a diameter of 0.5 mm, a volume of 0.065 mm^3, and a surface of 0.785 mm^2. Each liter of air can therefore generate a total surface area of 12 m^2. A patented rotating nozzle system cleans both the foam tube and the foam collector. A valve controls water that cleans the nozzle system. A regulation allows the skimmer to be operated under variable water column height.

The following types are available:

Type/Diameter	250	300	500	700	850	1000	1200	1500	2000	2500
Volume m3	0.07	0.10	0.27	0.63	1.40	1.90	2.71	4.30	7.50	11.70
Height m	2.3	2.3	2.3	2.6	3.6	3.6	3.6	3.6	3.6	3.6
Air m3/h	1.2	0.9	5.0	7.7	11.0	15.0	22.0	35.0	75.0	115.0
Power kW	0.35	0.35	0.75	0.85	1.40	2.80	3.60	4.20	11.40	14.55
Flow m3/h*	2	3	8	19	40	56	81	125	226	350

*The flow rate is calculated at an optimal retention time of 2 minutes.

HOLDING

Sand filters designed for easy operation

These high performance sand filters are made of glass-fiber reinforced polypropylene or polyester. The units are available in different sizes (0.4-0.9 m Ø) with or without a cross-shaped water-distribution system, and outlet valves. They include a manometer (with automatic and/or manual air bleed) and can be completed with pipes for immediate connection with 6-way backwash valve, self-priming (6-32 m³/h) filter pump (230 Volt/50 Hz) with large plastic debris collector, optionally mounted on a noise reducing plastic plate. Special filter sand (0.4-6.0 mm) is available as an accessory. An automatic backwash system, which guarantees safe and easy operation, is available for all models.

Pressure filters with backwash technic

This compact pressure filters with integrated UV-C lamp (for up to 9,000 hours) are complete systems with ecological threefold filtration technic (mechanical/biological/optical), to solve all filter problems in garden ponds or aquarium tanks. An integrated user friendly backwash technic allows an easy maintenance without opening of the filter. The water leaving the filter has the same pressure as the entering water and therefore may can also be used for waterfalls or injectors and therefore provides for a better oxygenation. The filters can stand free or set into the ground, below or above the water outlet and used for fresh- or saltwater. They are available in the following sizes:

Type	PA-4	PA-8	PA-12
Volume	25 l	30 l	35 l
Diameter	37 cm	37 cm	42 cm
Height	53 cm	61 cm	61 cm
Weight	7 kg	8 kg	10 kg
UV-C	9 W	11 W	18 W
Capacity	2.0 m3/h	300 m3/h	4.5 m3/h

The filters are equipped with a span-ring and 1" hose connections, for a recommended water flow of approx. 2.000-4.500 l/h and a tank or pond size of 2-6 resp. 4-12 m³ (with resp. without fish-stock).

HOLDING

Swirl separator with filter chambers

These new compact filters are used typically for holding units and garden ponds and are very effective for fine and raw filtration. The filter unit can be operated by gravity or by pumps and can be placed either in or on the ground.

An elongated inlet opening ensures that the water in the vortex rotates slowly so that the larger waste particles are trapped in the center bottom of the vortex tank. The built-in channels force the water to flow through the filter media (e.g.: filter brushes, Japanese mats, aqua rock, bio balls, polyfoam, etc.). Waste particles concentrate at each cone, where they can be removed routinely by opening a gate valve. The units are made from fiber-glass reinforced polyester, and are delivered with cover and outlet pipes. They are available with or without filter material.

The following Center-Vortex types are available:

Type	C20	C30	C50	C80	C100
Size L + W cm	85	107	140	165	193
Height cm	65	75	80	100	100
Vortex Ø cm	45	50	75	95	110
In/Outlet Ø mm	110/110	110/110	110/110	110/160	110/160
Flow capacity m3/h approx.	4	6	9	12	16
Pond volume m3 (max.)	20	30	50	80	100

Row-Vortex with 3-5 filter chambers, on request.

Biofilter media with large surface area

For recirculation and holding systems or the water purification and gas-stripping in fish farms, ponds and aquaria we deliver also just the bio-filter media.

This plastic material (polyethylene or polypropylene) can be used in flow-through- or trickling-bead filters, and has a large specific surface area ranging from 100 to 900 m^2/m^3, which allows a high rate of biological cleaning of ammonia and nitrate (nitrification/denitrification) of about 0.5 g NH_4-N/m^2 filter surface area (in freshwater at a temperature of more than 20 °C). In seawater (40-60 %) more filter material is recommended.

HOLDING

Heaters and chillers for liquids

These compact (CE-certified) heating and cooling systems can be used for a variety of liquid media (also seawater or foods) and are safe and secure in operation as there is no direct contact with the media. Electronic control using a high precision thermostat and the continuous display of temperature by a digital thermometer allows precise and easy temperature control. An environmentally approved refrigerant (R 134a or R410a) and low power consumption are features of these transportable and easy to install systems. The standard connection is 230 Volt 50-60 Hz. Optionally an integrated UV-C system is available.

The following standard types are available:

Type	TK-500	TK-1000	TK-2000	TK-3000	TK-6000
Power Watts	250/400	400/400	650/400	710/800	910/1200
Connections Ø mm	16/20	16/20	16/20	20/25	20/25
Weight kg	17	20	22	42	50
Dimensions cm	31x31x42	31x31x46	31x31x50	60x39x56	60x39x56
Capacity liter	500	1000	2000	3000	6000

Water disinfection with ultra-violet light

Ultraviolet light (UV) destroys micro-organisms by changing their genetic information (DNA), but does not produce residual or hazardous by-products, nor does it affect the taste, odor or color characteristics of the treated water.

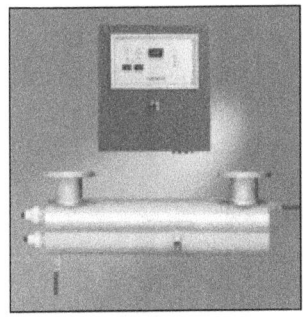

The hearts of these UV-systems are high-performance-lamps, which provide a stable UV output through a wider temperature range, and higher degrees of effectiveness and stability than other, conventional lamps. They have a high UV power output (up to three times that of competitive low pressure lamps) and long operating life (9,000 h). The lamps have a high UV emission at the effective wavelength (254 nm), which makes it possible to destroy up to 99.99 % of the pathogens in water.

By studying the UV intensity distribution in a range of reactor geometries and hydraulic characteristics, specific reactor types where developed to provide disinfection at flow rates up to several m^3/h. The control panels have electronic controllers containing smart chips that measure important operating parameters.

HOLDING

Polyester tanks in all shapes, sizes and colors

For more than 30 years these well-proven tanks have been constructed of layers of glass-fibre reinforced polyester (GRP) which is absolutely stable against climatic conditions or UV radiation. Their conical form, funnel-shaped outlet and absolutely smooth inside finish ensure the tanks are effectively self-cleaning, making it possible to hold sensitive bottom fish or fish larvae. The tanks and especially their rims are very strong, allowing the attachment of feeders and other equipment.

Circular tank   Square tank   Rectangular tank

Optional laminated feet make it possible to place the tanks safely on any surface, and ensure easy handling and transportation. The tanks are available in different shapes and sizes, and are available in all colors of the RAL-scale. The swing- or telescopic outlets are made from PVC pipe and allow regulation of water level. They are fixed under the tank bottom and project to the tank side. Bottom grids and stand sieves are available as accessories.

Circular pools with liner

These pools (2-8 m Ø) can be placed in or on the ground and consist of a strong, 0.4-0.7 mm galvanized iron wall which is painted inside and plastic-covered outside. The lower and upper bottom- and hand-rails are made of strong plastic to ensure a long life. The pools are lined with a strong, 0.6-0.8 mm black or blue PVC liner, which fits exact. Optional bottom protection pillows (5 mm) are available. Some of the available sizes:

Volume	Diameter	Height
2600 l	2.0 m	0.9 m
8000 l	3.0 m	1.2 m
18000 l	4.5 m	1.2 m
22000 l	5.0 m	1.5 m
36000 l	6.0 m	1.5 m
75000 l	8.0 m	1.5 m

We supply also just the strong, soft PVC pond liner material, 0.5 to 1.5 mm in thicknesses, in black, green or blue color, in rolls 2 to 8 m in width, and 15 to 50 m in length, or as finished liners in any size.

BREEDING

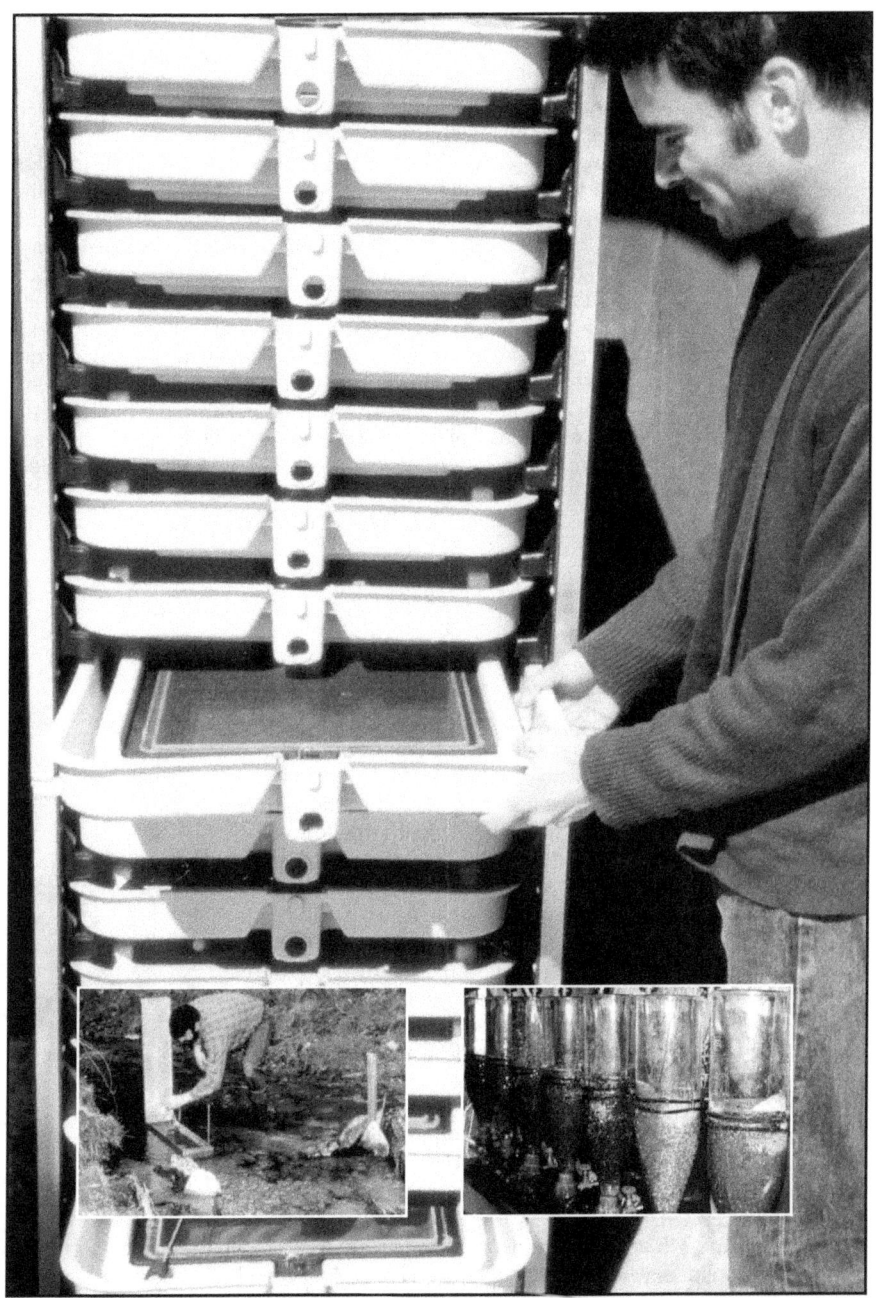

BREEDING

Vertical incubators to save space

These vertical incubators guarantee safe conditions for the breeding and hatching of salmonid eggs (like: salmon, trout, charr etc.) and can be combined and arranged in different ways, to utilize the available space as effectively as ever possible.

The water (min. 6 liter/minute) flows from the inlet, through the trays on which the eggs are placed, and exits over the front and flows via the side canals to the tray below, and so on, so that all trays in the stack will be supplied with sufficient water. Each tray can be removed for examination without disturbing the other trays. They are available in units of 4 to 16 trays, (for about 50,000 to 200,000 eggs) with a footprint of approx. 60 x 60 cm. The trays are made from strong, non-toxic plastic material (approx. 50 x 40 x 4 cm) with PVC coated polyester screens. The aluminum frames are included. Optional isolation panels for the front and back (clear or black), and tray segregation baskets are available.

Type/Trays	4	8	12	16
Egg capacity liters	4	8	12	16
Height cm	44	86	137	176
Weight kg	23	45	68	90

Vertical incubators to save water

This vertical incubator allows optimal conditions for the incubation and hatching of salmonid eggs (like: trout, charr, grayling etc.) and save water. The water (only 1.8 l/min) from the top tank with float switch and filter tube flows by gravity trough the trays, from the central inlet funnel through the sieve baskets where the eggs stay, and leaves the tray via the outlet holes all around the sides, where if lows down the cone shaped tray to the next one a. s. o., so that all trays will be supplied with sufficient oxygen. Without disturbing the other trays, each tray can be drawn out and controlled easily and provides optimal conditions from the fertilized egg to the swim-up stage. The corpus made of black plastic has a removable and roll-up front cover also made of plastic material, which reduces light within the incubator. The incubator is available with 10 trays and baskets (for 100.000 eggs respectively up to 1 liter eggs per tray) made from aluminum (approx. 40 cm diameter and 5 cm high). The whole incubator has dimensions (W x D x H) of 48 x 62 x 128 cm.

A support frame (45 cm high) made of stainless steel is available as accessory. The system can optionally be equipped with a chiller or operated in recirculation.

BREEDING

Breeding troughs with trays

For easy incubation with a good overview these breeding troughs are used worldwide for Salmonid eggs like: salmon, trout, charr, and graylings etc.

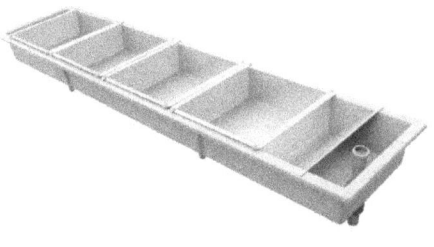

The troughs, made of high quality glass fibre reinforced polyester, are completely smooth on the inner sides and therefore easy to clean and disinfect. They are available in two sizes (length 215 or 360 cm) holding up to 4 or 7 incubation trays respectively. They are about 40 cm wide and 17 cm tall. The trays (approx. 40 x 40 cm) fit exactly.

Because of a new production technique, the stainless steel screens (with 1-2 mm holes), are inserted perfectly into the trays, and allow a constant exchange of water. Each tray has an incubation capacity of about 10,000 to 20,000 eggs (1-2 liters). To use the trough for first-feeding after hatching, a screen is available as an accessory.

Breeding boxes for running waters

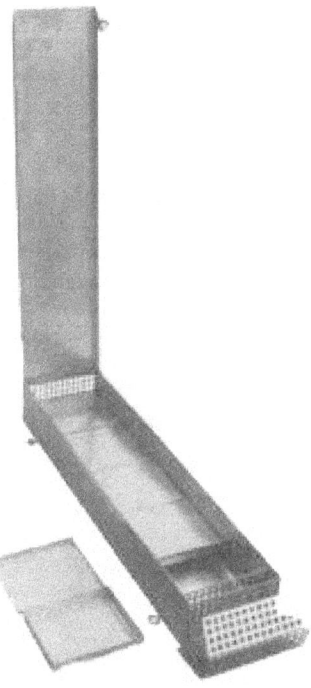

An experience of many years in egg incubation and fry rearing, as well as stocking experiments in various waters, where the base of this new invention for the hatching of salmonid eggs in natural flowing waters. The flood save and closeable box is made of stainless steel with dimensions of (L x W x H) 85 x 20 x 12 cm, weight approx. 8 kg). It is used for the controlled incubation and hatching of salmonid eggs as well as for the breeding and rearing of fry for (re)stocking purposes. The box is simply set into the water (front flap against the current and cover a little above the water line) and anchored by the eyes to the bottom ground. Depending on fish species and egg size, the box supports a save environment for up to 10.000 eggs and fry until exogenous feeding. The eggs lay in baskets, which can be fitted with special hatching substrate. Under "normal" conditions maintenance is limited to the periodical control (1 x per week) of the system, where results of 90-95 % are reached in average. When the fry reach the stage of exogenous feeding they can distribute themselves in the brook or river.

BREEDING

Hatching jars made of plexiglass

Used successfully for over 20 years by research and commercial hatcheries and fish producers, here and abroad, this McDonald type hatching jar become the standard of the fish farming industry. The jars can be used for a wide variety of species: carp, catfish, burbot, pike, perch, sturgeon, etc.

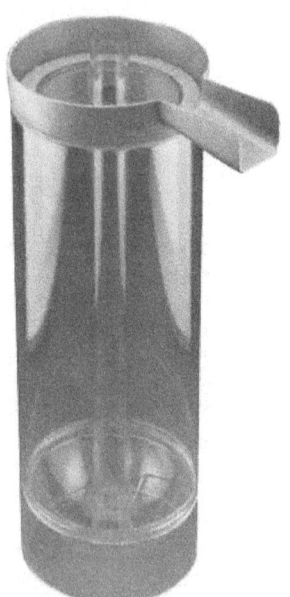

Made of high impact, non-corrosive, easy to clean, transparent plastic (Plexiglas), this uniquely designed hatching system enables uniformly distributed water to rotate the fish eggs gently and evenly. It works very simply: water from the supply line is directed down the open-ended delivery tube in the center of the jar. The tube (3 cm Ø) easily accommodates the required water flow so that the eggs at the base are rotated uniformly and evenly. The jars are 46 cm tall, 16 cm in diameter, and have a volume of nearly 7 liters (for up to 3.5 l eggs). Each weighs only 1.1 kg and is supplied complete with delivery tube and screen. The circular filter/mesh screen prevents eggs from spilling out of the jar, while the large spout (length 7.5 cm, width 5.3 cm) at the top lets hatched fish swim out at just the right time.

Now you can virtually duplicate nature's way of hatching eggs with this durable, see-through incubator jar. So next time you want your eggs "over easy", make sure you are using this hatching jar.

Incubation silos made of plastic

For breeding of larger quantities of fish eggs (like salmonids), normally from fertilization to hatching, this special silos made of plastic have been proven. The cylindrical egg container has a volume of approx. 30-60 l (diameter 33-48 cm, height 65-90 cm) and is delivered complete with in- and overflow, as well as screen and stand.

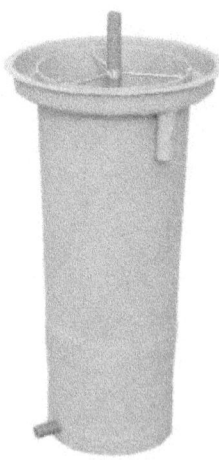

For holding of larger quantities of fish larvae (like cyprinids), normally from hatching to feeding, this special silos made of polyester are available. These conical larval containers have a volume of approx. 90-240 l (diameter 65-87 cm, height 113 cm) and are delivered complete with in- and overflow, as well as screen and 3 feet made of stainless steel.

BREEDING

Incubation jar system with overflow

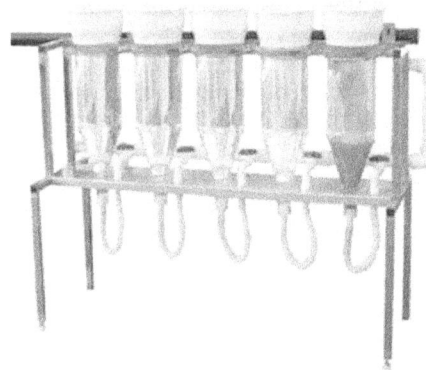

These simple and cheap jar systems, developed in Eastern Europe, are made of stainless steel frames and are equipped with "Weiss" jars made of glass which can be serviced and supplied separately. Each jar with a volume of about 8 liters and bottom sieve, has its own inlet pipe with valve and an overflow with pipe. The eggs of different fish species, like carp, pike, perch, catfish, burbot, sturgeon etc.) are incubated in the jars and rotated gently by water flow. After hatching the larvae usually leave the jars by themselves with the water flow or are removed from the jar by the farmer.
The system can be built with any number of jars (up to 10 pieces in single and up to 20 pieces in double row). Incubation jars with lamp and compressor for hatching of eggs of brine shrimp (Artemia), on request.

Breeding jar system with top tank

These well designed jar systems, developed in Western Europe, are made of stainless steel and are equipped with "Zougg" jars made of glass which can be removed and serviced individually. Through the inlet tank on top, which allows to degas the water, all jars can be individually supplied with water. The tank and feet are somewhat adjustable in height. The eggs of various fish species (cyprinids, esocids, thymallids, coregonids etc.) are incubated in the jars and rotated gently by the water flow. A breeding cone for salmonid eggs and an overflow sieve (1-2 mm holes) to avoid eggs or larvae being spilled out by water, is available as accessory. The system can optionally be equipped with a heating/cooling system and operated in recirculation.

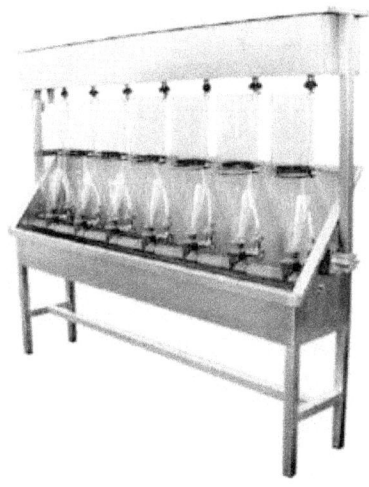

The standard systems have dimensions (L x W x H) of 80-200 x 40 x 165 cm and are equipped with 3-7 glass jars each with a volume of 8 liters. Other dimensions or jar volumes are available on request. Single glass jars, with or without a stainless steel support-stand, are also available.

BREEDING

Hatching substrate for vital fry

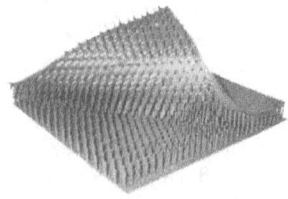

Mainly for salmons, but also for many other salmonids, this 2-layer hatching substrate (40 x 40 cm) have been used and proven well. The eggs are incubated on the green or grey hatching substrate as usually practiced. After hatching the larvae move through the spaces (about 2 x 15 mm) and shelter in the substrate nipples of the soft rubber like plastic substrate (3 cm high).

The hatching substrate imitates the gravel substrate naturally occurring in the brooks or rivers and provides optimal developmental conditions for the fish larvae. The fry are therefore stronger and healthier, they also have less deformations and lower mortalities, and growth improves (up to 10-20 %).

Manual device for sorting salmonid eggs

First developed at the Austrian Institute of Fisheries Economy, this handheld egg-sucker has been redesigned, allowing quick and easy removal of salmonid eggs. The pistol-shaped device is held in one hand, while the other hand holds the suction ball. This has to be squeezed until the unit overflows at the end of the tube. Now, each dead egg can be approached with the suction pipe and when the button on the handle is pressed, the egg will be sucked into the plexiglas collecting unit. With some experience, it is possible to work quick and precisely.

Automatic machine for sorting salmonid eggs

This egg-sorting machine uses the most current technology to sort 100,000 eggs per hour. It is available with or without counters for both live and dead eggs. Its dimensions (L x W x H) are approx. 78 x 35 x 27 cm, and it weighs 23.5 kg. The egg tank is made of clear and white plastic with a capacity of 19 liters. It has been used around the world and will accommodate any local voltage, which is reduced to 12 V DC in the interest of safety.

Using the patented standard disk, the egg sorter will sort salmonid eggs ranging from 5,000 to 18,000 per kg, or if equipped with the salmon disk, it will sort any size salmon egg. Best of all, you do not have to change disks to handle different sizes of eggs. The patented technology uses fiber-optics and modulated infrared light to scan the eggs, ensuring the highest level of accuracy (approx. 98-99 %).

FEEDING

FEEDING

Automatic feeders with solar panel

"Solaris" feeders are designed and constructed using the most modern components and technology to meet the current requirements of aquaculture economics.

The "INTERVAL" model is delivered with a photo-electric panel, which generates sufficient energy to drive the motor directly. It is equipped with a computer and accumulator (to accommodate bad-weather days), which are mounted in a splash-proof box (directly on the feeder) below the solar panel. The unit can be adjusted to exact feeding intervals, feed amount, and time. The feeding amount (of 1-5 mm pellets) is adjustable from 0.1 kg up to 100.0 kg/day.
The "SUPERSPREADER" model is additionally equipped with an adjustable sector- and/or round spreader. The radius is adjustable from approx. 2 to 8 m, (and up to 20 m on request), and is delivered with a larger solar panel and accumulator.
The "UNIVERSAL" model is additionally equipped with a larger outlet, and can be used for flake feeds, seeds, and pellets up to 12 mm in size.

The white or green feed hoppers are available in five sizes (15, 25, 45, 75 or 95 liters). As accessories, tube- or ring holder hopper supports are available. Special supports on request.
All models can be supplied without the photo-electric panel, battery and timer, for connection to a central power supply (220-240 V, AC) with an available transformer (12-24 V, DC), control box, and timer (for 1 or up to 8 and 24 feeders).

Electric feeder for fry

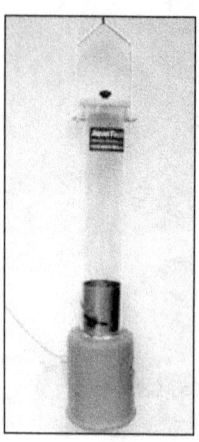

The "HATCHERY" model is also available for connection to this central timer and distributor box. It is a disc-feeder with a 2.5 l feed cylinder developed for delivering fine fry-food (0.1-2.0 mm) in the hatchery. The feeder is suspended above the fry tanks. Power is provided by a control box equipped with a 220V AC/12V DC) transformer and a timer. The beginning (i.e. morning) and the end (i.e. evening) of the feeding period is set on the timer. The distributor box automatically sends a pulse that simultaneously switches on the motors of up to 24 connected feeders for a short time to move the conveyor disks. The interval between pulses is infinitely adjustable from 0.1 second to 10.0 hours. To regulate the amount of feed delivered at individual feeders, the slot between the conveyor disk and the plexiglas cylinder can be adjusted by slackening the adjustment screw. Individual timers are possible optionally.

FEEDING

Automatic feeders with integrated battery

This independent feeders where developed in Austria and are successfully used since more than 20 years in small to large fish farms.

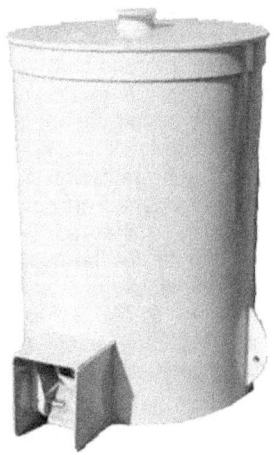

The user friendly feeding system with patented mechanism (spindle) allows easy feeding and supply of food to day- and night active fish or other animals. They can be powered by usual (Mono D) batteries (3 to 12 V/DC), which are supplied with the feeder. Because of the optimized system, the batteries have a long self-live, depending on feeding times and durance, usually about one season (often also over 1 year). The timer clock with battery tester, which is integrated in the hopper, allows exact settings of feeding times and pauses. The feed supply can be regulated by a turn-able knob and ranges from about 1 to 500 g per hour. Standard particle or pellet sizes of feeds can range from 1 to 5 mm, smaller or larger food particles (also corn or other seeds) are possible optionally. The robust, grey plastic containers (made from polypropylene) guarantee for a long feeder life.

The standard feeders range from hoppers with a volume of 10 to 260 liter (larger hoppers are possible on request). The larger feeders (from 40 liter volume onward) are also available with an integrated ejector-mechanism, which ejects the dry-food 3-6 m (optionally up to 8-12 m) in front of the feeder. The feeder is usually placed on the edge of a tank or above a dock and fixed via the two mounting holes on the sides. Optionally photovoltaic panels are available. All feeders are also available without batteries, and with or without timer, for the connection to an electric network with transformer (220 V/AC to 12 or 24 V/DC).

Pneumatic feeders with air blower

This easy to use and very robust pneumatic feeders are specially recommended for feeding 1-6 mm pellets on ponds or raceways, where large feeding areas (up to 10 m^2) and a wide feed distribution (up to 12 m) are necessary. The compact and lightweight container made from stainless steel is suitable for the rough conditions in the field. It has a capacity up to 60-180 kg of dry food, has 2 extendable pipe outlets with distributor plates and contains also the blower (220 V/50 Hz, 250-375 W) and a timer etc. The timer allow to set up to 4 feeding times per hour, and the feeding time can be adjusted independently. The feed quantity and be additionally changed at the 2 pipe outlets by a screw. The whole feeders are 55-85 cm in diameter and 83-100 cm tall.

FEEDING

Belfeeders with clockwork spring

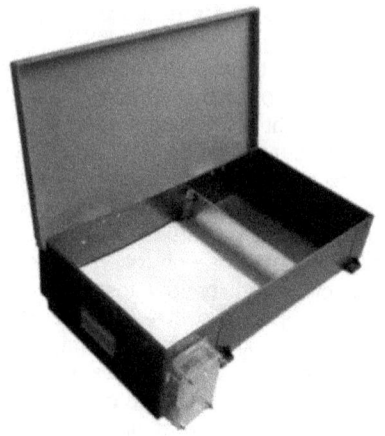

For more than 30 years these feeders have been used for safe and exact feeding of fish fry and fingerlings. The clockwork runs for up to 12 or 24 hours, and is started when the feed-belt is re-wound (in the morning). Up to 3-5 kg of feed per cycle can be placed on the belt for delivery. The new clockwork has a strong, corrosion-resistant steel spring and a splash-proof acrylic cover. The plastic box is weather-resistant, and ensuring a long life for the feeder. The feeders are available in 2 sizes measuring (L x W x H): 560 x 200 or 300 x 150 mm, and can be used at ponds or tanks that lack electricity.

Automatic feeding device for frozen zooplankton

Twenty years of worldwide experience in the field of plankton research and harvesting, and larval feeding was the basis for the new "plankton feeder", which was developed in cooperation with the Institute of Fish Research.

This device is designed for the professional use of frozen zooplankton as a larval diet in marine- and fresh water hatcheries for the production of high quality fingerlings like bass, bream, grouper, cobia, turbot and other species. The technique is based on a timed "wash down" of layers of feed from large blocks of frozen plankton (up to 20 kg), and the homogeneous distribution of the thawed plankton to the larval tanks by a specially developed self-cleaning distributor. Thawing and "wash down" of the food layers is performed either by spraying or by flooding, the remaining block of plankton is then re-frozen.

The particle size of the feed can be adjusted by a sieving unit. In order to avoid membrane damage and nutrient leaching from the thawed plankton organisms, the feed is delivered to the larval tanks (up to 16 or 32) by gravity flow only.

AERATING

AERATING

Injectors for aeration, circulation, oxygenation and de-stratification

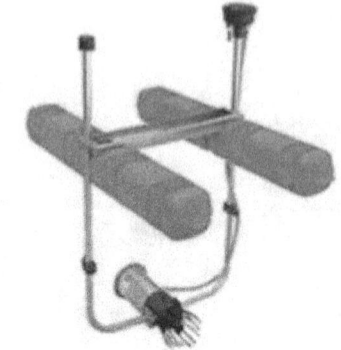

These proven injectors are one of the best and most advanced aeration systems worldwide, and also suitable for large ponds.
The injectors work with a submerged, adjustable, maintenance-free motor for continuous trouble-free operation. The rotating propeller (2800 r.p.m.) produces a current, which draws air from the surface in the form of fine, evenly distributed bubbles directed toward the bottom of the pond. Aeration is up to the surface and the circulation so created displaces the surrounding water, thus breaking up any stratification. This constant circulation forces the water in the pond to turn over, bringing cooler bottom water to the surface, where it also picks up atmospheric oxygen. The dissolved oxygen content is as high as possible.
Usually there is an oxygen deficit at depth and since the depth of the plume can be adjusted, oxygen can be delivered where it is needed, not simply at the surface. Depending on the installation, depth and current, sediments can also be removed. If the aerator is positioned near the water surface it can also act as an ice-clearing machine in winter. Strong circulation also eliminates organic matter deposition. Finally due to the flotation phenomenon (foaming), these aerators make it possible to remove excess algal growth, colloidal substances, mud and other suspended particles, and improve water quality. Depending on power and depth, the air flow can be up to 35 m^3/h. At an initial oxygen concentration of 6 mg/l, a single 1.0 HP unit, can introduce over 0.32 kg O^2/h, which is the oxygen required by more than 2600 kg trout (each 250 g) at 10 °C, or for as much as 3000 kg of other species at 20 °C. The aerators are supplied complete with motor (0.75 kW/1.0 HP, 230 V/50 Hz, or 1.2 kW/1.5 HP, 380 V/50 Hz), support, floats, and propeller protection basket. They units are compact, lightweight (20-22 kg), and very easy to install.

Injector that requires no additional energy consumption

This "ECO" injector is a static aerator that can be fitted to a 1" (2.5 cm) water inlet pipe to tanks, raceways and small ponds. It offers a cost-effective and efficient use of an available piped water source for degassing, aeration and oxygenation. It requires no additional energy. In enriching the water with air or oxygen, it can create good circulation with less water than is usually necessary.
It works as a venturi and has no moving parts, by exploiting the action of the water passing through it at high speed. The effect is achieved by using suitable flow rates and pressures (from 0.2 bar and 0.4 l/s on) with one of the 3 interchangeable nozzles (for flows of 30, 50 or 100 l/min) which create a pressure drop and consequently a tremendous suction capacity.

AERATING

Propeller aerator with mushroom spray pattern

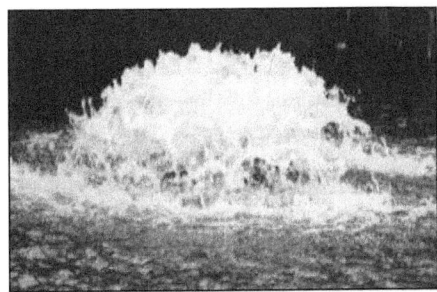

This propeller aerator was developed for intensive production of fish in tanks, raceways and ponds. The maintenance-free, heavy duty, motor (230 or 380 Volt) has a mounted propeller that optimizes water circulation, and has an oxygen transfer rate of approx. 1.5 kg/kWh. The small float size (70 x 70 cm) and the unit's light weight make it easy to install.

The following types are available:

Type/Power kW/HP	0.15/0.2	0.37/0.5	0.75/1.0
Water circulation m3/h	50	100	180
Water splash Ø cm	160	180	250
Water splash height cm	60	75	90
Weight kg	26	28	30

Accessory protection baskets (with 4, 7 or 12 mm grating) are available, which protect smaller fish from injuries. Additionally, conveyers for drawing water from deeper in the pond can be built to specified lengths.

Aerator pump with fountain spray pattern

This "Spray" aerator pump is especially suitable for ponds, since it introduces a lot of oxygen into the water. If the floating pump is fitted with the spray nozzle it becomes an aerator with fountain spray pattern. If the spray nozzle is replaced with the pipe adaptor (also supplied) it can be used as a floating pump for irrigation. Changing these two functions does not need any tools. The pumps are reasonably priced and are easy to install, either floating on the water surface or submerged to any depth. The following types are available:

Type/Watts	250	550	750	1100	1500	1800
Voltage (50 Hz)	220	220	220	380	380	380
Circulation m3/h	10	20	25	30	40	75
Splash height m	3-4	4-5	6-7	6-7	6-7	3-4
Splash Ø m	3-4	4-6	6-8	8-10	9-11	10-12
Pipe Ø mm	50	65	80	80	80	100
Weight kg	8	14	15	15	16	20

AERATING

Paddlewheel aerators for intensive production

Paddlewheel aerators are used all over the world, especially in intensive ponds and raceways. They are available with 2-6 paddle wheels, in different power ratings from 0.75 to 2.20 kW (220 or 380 Volt) and weigh 80-120 kg. Mass oxygen introduction is usually about 1.6 kg/kWh. Wheel and frame are made of steel, the paddles from special plastic material and the H-formed floats from resistant plastic.

Paddlewheel systems for effective oxygenation

Aeration and/or oxygenation are especially of economic importance in intensive farms. At these completely new developed systems all possible failures and defects of other known products have been removed. The systems have been tested under hardest conditions and have a very high efficiency, with low operational costs at the same time. There are no wheel bearings (approx. 140 r.p.m) necessary, and also no couplings and discs, this reduces friction loss and increases the efficiency, and also reduces the number of parts that could become defective with time. Long life and high efficiency are also guaranteed through the worm-gearing. The motor is specially sealed against moisture. Paddles, wheel and frame are made of stainless steel, and the floats from thick UV- and ozone resistant material. The rate of oxygen introduction is usually about 1.6 kg/kWh. Paddlewheel Aerators are available in different power ratings (180, 250, 370, 550 and 750 Watt) for standard 380 V (230 V on request) and measure approx. 1.5 x 1.6 x 0.6 m (L x W x H) and weigh 30-70 kg.

Paddlewheel aerator

Oxygen cover

To allow the use of these paddlewheel aerators with pure oxygen instead of air, an oxygen cover is available as an accessory. This specially shaped cover guarantees that no oxygen (feed up to 15 l/min) disperses to the surface after the water leaves the aerator. As there are no seals necessary, this also guarantees that there is no oxygen wasted, thus making the system very (up to 90 %) efficient. Optionally, an oxygen-return-system is available, which allows to reuse excess oxygen.

AERATING

Ceramic plate diffusers for micro gas bubbles

This diffuser is a robustly engineered product with high specification and performance, and is perhaps the best and most efficient diffuser available worldwide to date.

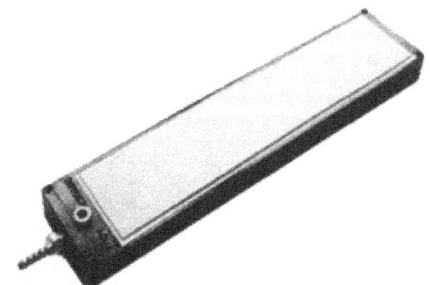

The ceramic plate has been designed to be very strong, its thickness (approx. 12 mm) is greater than that of more conventional plates. The pore size is only 0.3 microns, which produces a cloud of tiny gas bubbles (10-200 microns) in the water. Depending on the gas (oxygen or ozone) and water chemistry, the diffused gas may pass directly into solution with no bubbles.

The oxygen transfer efficiency can approach up to 100 % passing into solution, and the oxygen saturation in the water may exceed 120 %. The ceramic flat plate diffuser gives a very small bubble size, and can be used wherever a small high rate diffuser with high oxygen transfer efficiency is required, like in transport tanks or production units. The working gas pressure is 25-35 psi. Do not exceed 50 psi (3.5 bars) as it may result in damage to the plate. The diffuser weighs 3.55 kg, outside dimensions are 450 x 110 x 28 mm with a ceramic plate of 400 x 90 x 12 mm (360 cm^2) for a gas flow of up to 6-8 l/min. It has a 6 mm oxygen hose adapter.

Rubber diffusers with stainless steel frame

These extremely durable and effective diffusers are suitable for use during live fish transport and in emergency, and can be used with pure oxygen. There are approx. 1000 pores/meter, arranged in 6 rows. The special oxygen hose opens only when oxygen is supplied and closes again if the supply stops. Thus, pore blockage is prevented and the diffuser is always ready to operate. Uniform bubble size due to the straight hose mounted in a stainless steel frame is a feature.

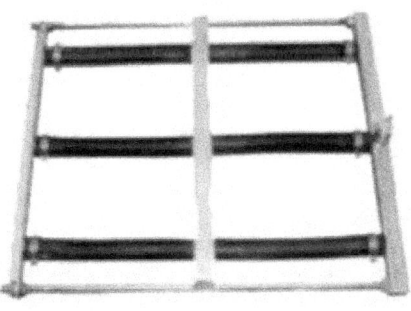

The diffusers are available in different sizes (40-70 x 50-150 cm). Other diffuser sizes and shapes on request. The 6 mm sturdy hose connection is standard. For those who wish to build their own units we supply diffuser hose (19 mm i.d.) and pressure hose (6 mm i.d.) by the meter, as well as end pieces and connectors.

AERATING

Professional ceramic diffusers for air

These fine-pored ceramic tubes allow the diffusing of air in fine bubbles. The diffusers should be installed at the same depth, as deeply as possible below the water surface. The floating cross-shaped diffuser is placed on the water surface in the center of the tank or pond and fixed with a rope to the shore so that it moves a little and aerate a larger area.

The cross-shaped diffusers are available complete with 20 mm hose tail (other sizes on request) and can be delivered with or without float. The 1 m long supply tube is available in shorter lengths. Each ceramic diffuser tube is 18 cm long and 6 cm in diameter, and can introduce approx. 1.5 to 2.0 m^3 air/hour, (6-8 m^3 per cross diffuser). They are also available as self-sinking single tube diffusers with hose tail. Supply hoses (clear) and air distributors with valves are available as accessories.

High performance gas diffusers

These are among the most robust and versatile fine bubble diffusers available. They resolve many of the problems associated with other diffusers and have proved themselves over 10 years of use in many industrial and waste-water treatment applications. The diffuser will deliver large quantities of air, oxygen or carbon dioxide with a low pressure drop and small bubble size (approx. 1-4 mm). Fine bubble diffusers are inherently more effective than coarse bubble diffusers in providing a greater mixing action and aeration efficiency. The diffusers are of a tubular semi-flexible construction (32 mm Ø).

An outer polyester sleeve encases silica particle ballast and an inner nylon distributor. The internal ballast is not fused together, therefore precipitates such as iron and carbonates tend to break off, and can be flushed out. The internal ballast (weight approx. 1.6 kg/m) means that the diffuser does not need to be anchored to the tank bottom. The unique design of the diffusers makes them resistant to biological fouling.

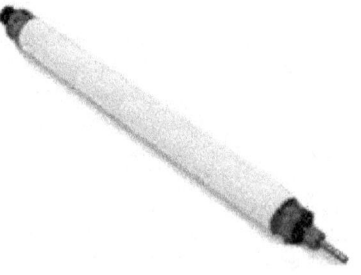

The diffusers are sized (up to 3 m in length and can be extended) according to the flow (m^3/h) required. A diffuser 33 cm in length will deliver approx. 1 m^3/h air or 1 liter/min oxygen with a pressure drop of less than 1 psi. All diffusers are supplied with either 3/8", or 1/2" hose tails in acetyl plastic.

AERATING

Side-channel blowers for water aeration or delivery

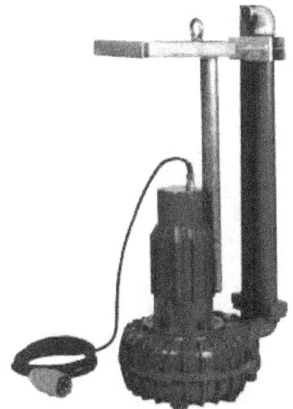

The design of these blowers is based on long experience in blower production.
The motor (standard 380 Volt/50 Hz, 2850 RPM) is made for endurance and is offered in different power sizes. The flow capacity based on air at 981 mbar and 20 °C (at the inlet connection) ranges from 80-500 m^3/h. Because of the small dimensions and compact construction, as well as the low noise level (64-72 dB at 1 m distance), these blowers can be placed almost anywhere. The blowers can be installed singly, in parallel, or in rows for an increase in pressure. On request the blowers can be produced for us in seawater, or for other voltages and frequencies. Air filters, pressure valves and support stands are available as accessories.

The typical emerged blowers (type V) are installed only on the land, but at almost any location. An advantage in recirculation systems is that the heat in the air of a room can be used to heat the water.

The special submerged blowers (type T) are installed directly in the water and therefore allow shorter piping, as well as lower pressure losses. They are much quieter and also save space. An additional advantage in recirculation systems is that the heat generated by the blower motor will heat water.

The following emerged blowers are available:

Type	V01	V11	V12	V22	V24
Flow capacity m3/h	80	100	200	300	500
Pressure difference mbar	75	100	120	160	200
Motor power kW	0.3	0.5	1.1	2.2	4.0
Weight kg	9	12	16	18	22

The following submerged blowers are available:

Type	T01	T11	T12	T22	T24
Flow capacity m3/h	80	100	200	300	500
Pressure difference mbar	110	150	150	200	200
Motor power kW	0.3	0.5	1.1	2.2	4.0
Weight kg	22	26	32	38	58

AERATING

Regenerative blowers for pressure or vacuum generation

These blowers summarize modern know-how in pressure- and vacuum generation. The motor is constructed for endurance. Because of the compact design and the low noise level, these blowers are easy install to anywhere. Lateral channel blowers work regeneratively. The casing and radial blade impeller forms a circular body. The medium (air) is accelerated by centrifugal force in the blade-channels and compressed. This process continues progressively, increasing the pressure until the air leaves the casing through the outlet. As accessories air filters, sound absorbers, and non-return valves are available.

Types with the following performance data* are available as standard:

P mbar	0	50	100	150	200	250	300	350	400	450	500	550	600
Single stage (kw)	m3/h	m3/h	m3/h	m3/h	m3/h	m3/h	m3/h	m3/h	m3/h	m3/h	m3/h	m3/h	m3/h
06-SH (0.20)	55	29											
V3-SH (0.37)	70	50	31										
30-SH (0.75-1.50)	124	109	94	78	64	49							
40-SH (0.75-2.20)	204	172	145	120	99	80	64						
45-SH (1.10-3.00)	249	215	182	148	114	78	37						
50-SH (2.20-5.50)	319	286	256	230	204	180	156	113	110				
60-SH (2.20-7.50)	393	362	335	308	283	260	235	210	185	160			
65-SH (2.20-5.50)	527	468	414	365	316	266	218						
70-SH (3.00-7.50)	522	477	433	389	350	310	273	235					
80-SH (4.00-15.0)	834	778	714	655	602	550	497	445	393	341			
90-SH (5.50-15.0)	1084	990	900	820	743	667	592	515	440				
Double stage (kw)	m3/h	m3/h	m3/h	m3/h	m3/h	m3/h	m3/h	m3/h	m3/h	m3/h	m3/h	m3/h	m3/h
10-DH (0.37)	30	23	15	9	2								
15-DH (0.55)	50	44	37	31	25	20	14						
20-DH (0.37-1.10)	72	62	51	43	35	28	22	17					
30-DH (0.75-1.50)	103	90	79	70	61	52	43	35	26				
40-DH (0.75-3.00)	144	131	118	106	95	84	73	64	55	46	37		
50-DH (2.20-4.00)	182	168	149	137	127	118	110	103	97	90	84	78	
65-DH (2.20-4.00)	269	242	220	201	184	170	156	143	127	112			
70-DH (2.20-7.50)	276	259	240	226	215	202	193	183	173	164	154	145	136
80-DH (4.00-9.00)	430	410	387	371	355	341	326	313	300	287	273	261	250
90-DH (4.00-11.0)	546	510	495	472	450	430	410	391	372	353	333	315	300

*At pressure with 50 Hz (= 2900 RPM) for air at 15 °C and 1013 mbar, tolerance +/- 10 %.

AERATING

Oxygen generators for intensive production

Several studies have shown adding dissolved oxygen into water on fish farms has a very positive effect on growth rates, feed conversion and mortality rates, factors that will affect the profitability significantly. On-site oxygen generators allow an uninterrupted supply of gas with a high purity output (adjustable 85-95 %, outlet pressure 4 bar). This means that you can produce gas where and when you need it, and in the exact quantity and quality you need, to reach gas self-sufficiency and independence from external gas suppliers. This oxygen generators are known as reliable, easy to use and low operational cost. They save energy and money as supply of oxygen cylinders or rent of liquid oxygen tanks becomes irrelevant. The average oxygen production cost are about 1.1 kWh per cubic meter of oxygen. In comparison with cylinders it gives overall cost reduction up to 80 %. We use only high quality components and all systems are designed for continuous operation.

This proved generators are based on the well-known PSA (Pressure Swing Adsorption) technology, using two pressurized columns with molecular sieves to ensure continuous production. Dry compressed air is blown through a valve to the vessel where the pressure is built to reach 5 to 7 bar (g). Oxygen is tied to a molecular sieve during the building of pressure and the nitrogen is allowed to pass through to the accumulation tank. While pressure is built in vessel, the second remains without pressure. A part of the produced gas is used for regeneration of the molecular sieve, which, in the case of Nitrogen, is a Carbon Molecular Sieve (CMS). The entire PSA system can be subdivided into several components like: air compressor, air dryer, air tank, oxygen generator, oxygen tank and accessories. Oxygen cylinder filling stations, on request.

Type	O2 kg/h	O2 m3/h	Consumption*	Tanks	Weight
OM-020	2.4	1.7	2.7 kW	90 l	354 kg
OM-040	3.5	2.5	3.6 kW	90 l	384 kg
OM-060	5.0	3.5	4.8 kW	90 l	446 kg
OM-070	6.7	4,7	6.5 kW	150 l	653 kg
OM-102	10.5	7.4	9.1 kW	320 l	836 kg
OM-170	14.9	10.5	10.7 kW	320 l	1086 kg
OM-230	21.4	15.1	14.3 kW	470 l	1416 kg
OM-234	28.5	20.1	21.0 kW	470 l	1455 kg
OM-333	37.6	26.5	23.7 kW	750 l	2220 kg
OM-335	44.2	31.1	28.4 kW	750 l	2312 kg
OM-503	67.0	47.2	42.1 kW	1000 l	3548 kg
OM-600	84.1	59.2	60.0 kW	1500 l	4605 kg

Capacity (±5 %) at 15 °C, 981 mbar, inlet pressure 7 bar, O2 purity 90 % (±1 %).
* incl. compressor, dryer and generator.

MONITORING

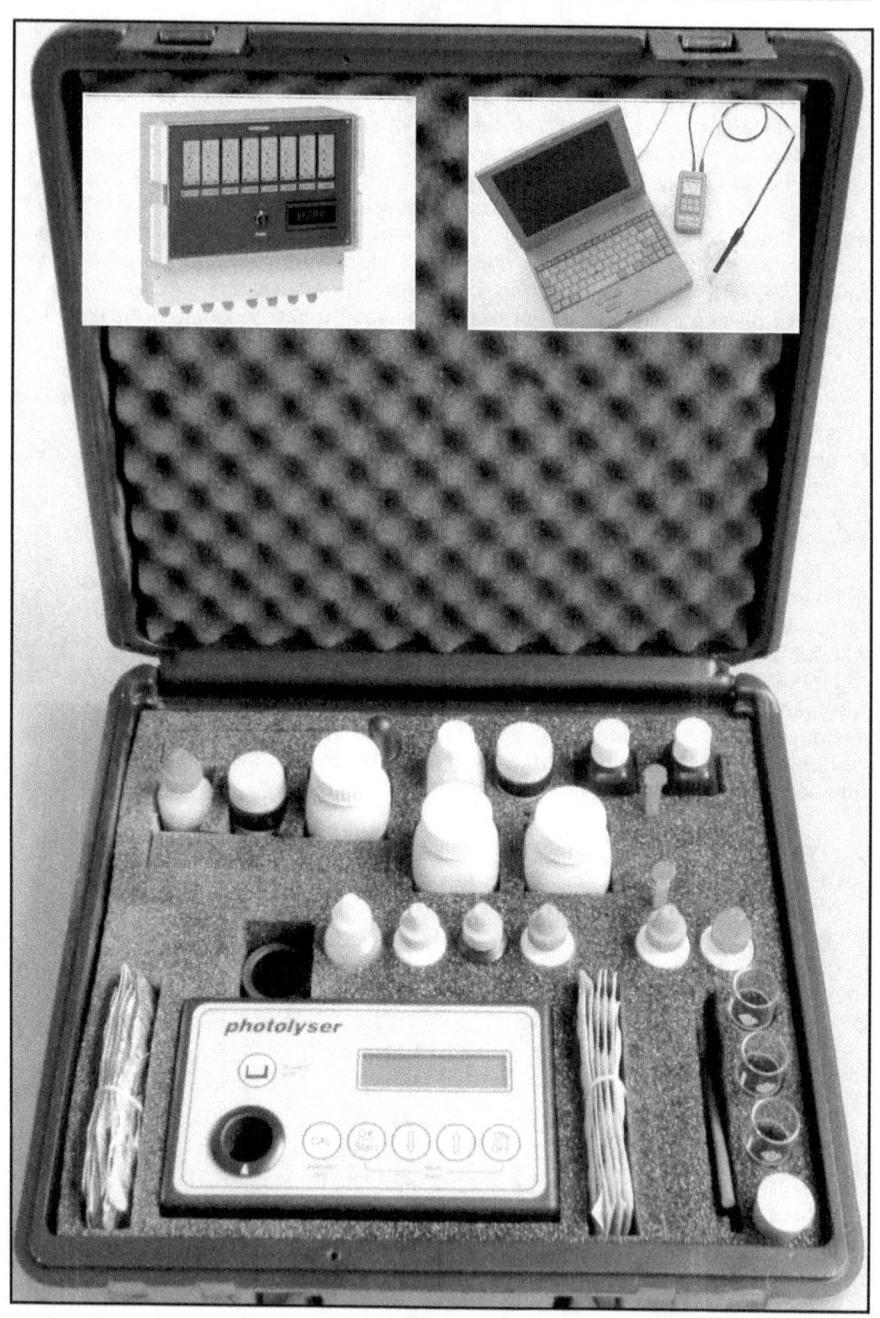

MONITORING

Microprocessor meters for professional monitoring

This improved meters were specially adapted for use in aquaculture and limnology. The compact (L x W x H 165 x 83 x 34 mm, approx. 305 g) splash-proof cover (IP 54), the plastic sensor (14 mm Ø, 150 mm long, with 1.5 m cable), the very long sensor life (approx. 2 years standby time), and the delivery with NiMH-batteries (2 x AA Mignon) and USB-cable with charger in transport case, make these meters perfect for field work or laboratory use. The large graphic multi-display makes the readings available clearly and quickly. The short polarization time (approx. 30 sec. max. 5 min.), the low flow current sensitivity (1 % at 5 cm/sec.), the high precision +/-1 digit (0.1 mg, 0.5 % or °C), as well as salinity correction, automatic air-pressure and temperature compensation (in the range of 5-40 °C) makes working easy. Automatic calibration with the calibration chamber (included) is an additional feature. Options include stainless steel sensors, and other cable lengths available. To enable you assigning measured values geographically, these meters are available with or without GPS.

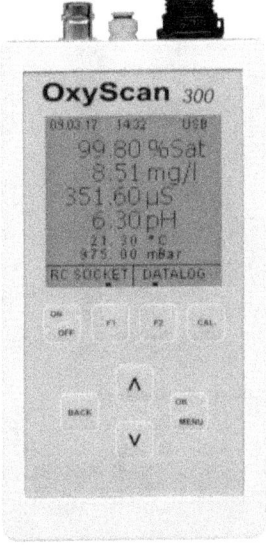

This modern devices have: an electronic compass, a data logger with adjustable start/stop time and measuring intervals between 5 sec. and 10 min., with automatic storage of all measured values (approx. 3 months during battery operation, also mains operation possible), a large internal memory (32000 locations), and an individual measured value storage on key press with 1000 memory locations. Additionally, radio controlled (RC) switching outputs for standard RC sockets and acoustic signals, for up to 4 RC sockets with separate two-point control for oxygen (absolute or relative), conductivity, temperature or air pressure, two additional alarm outputs, which trigger a standard RC bell when the set values are exceeded or subtracted, and two additional channels for optical and/or acoustic alarm on the meter. The power-saving graphics display (transflective with background lighting) with best readability even in direct sunlight, and adjustable contrast, activates automatically display and background lighting when the device is moved (can be switched off). The beep (alarm and key tone) volume and frequency is adjustable. PC-software for data visualization and data export included (with free online software update).

The model „Outdoor" has the following measurement: Oxygen 0,00-25,00 mg/l and 0,00-250,00 % sat (+/- 0,01 mg); Temperature 0,00-60,00 °C (+/- 0,1 °C); Air-pressure 50,00-1150,00 mbar (+/- 0,5 mbar).
Das model „Laboratory" has optionally the following additional measurement: pH-value 0,00-14,00 (+/- 0,01) or Redox-potential -500,00-+500,00 mV (+/- 0,3 mV). As accessories for hat model, an optical laser oxygen sensor with conductivity reading, as well as a connection box for up to 4 oxygen sensors is available.

MONITORING

Electronic meters for many water parameters

All these electronic devices measure 142 x 71 x 26 mm (L x W x D) and weigh approx. 155/255/300 g (incl. battery/sensor) depending on the model. They have a double LCD display (precision: ±1 digit) for measured values and temperature, min-/max-value memory, configurable automatic switch-off function (1-120 minutes), hold-function, foil push-buttons, automatic temperature compensation, 12 mm Ø sensors, serial connection, 9 V battery with warning and external energy connector, as well as integrated stand and hanging clip. Transformer and transport case are available as accessories.

The following meters are available:

GMH-33 Flow-Meter
- Measurement: Flow 0.05 - 5.00 m/sec;
- Options: Air humidity 0.0 – 100.0 % rH;
  Air temperature -40.0 – 120.0 °C;
- Outfit: Anemo-sensor with 5 m cable and connector.
  Air measurement as accessory. Telescope sensor on request.

GMH-34 Conductivity-Meter
- Measurement: Conductivity 0 - 2000 µS/cm;
  Salinity 0.0 - 70.0 g/kg;
  TDS: 0 - 1999 mg/l;
  Resistance 0.005 - 100.0 kOhm;
  Temperature 0.0 - 100.0 °C;
- Outfit: Graphite-sensor with 1 m cable fixed to the meter.
  Automatic cell correction.

GMH-35 pH-Meter
- Measurement: pH-value 0.00 - 14.00 pH;
  Redox (ORP) -1999 - +1999 mV;
  Hardness 0.0 - 70.0 rH;
- Options: Temperature 0.0 - 150.0 °C;
- Outfit: KCl-sensors with 1 m cable and connector.
  Automatic calibration.

GMH-36 Oxygen-Meter
- Measurement: Oxygen 0.0 - 60.0 mg/l (ppm), 0 - 600 % (sat);
  Air-Pressure 10 - 1200 hPa;
  Temperature 0.0 - 50.0 °C;
- Outfit: Galvanic-sensor with 4 m cable and connector.
  Automatic pressure and manual salinity compensation.

A PC-interface adaptor and software are available as accessories, which make an efficient and easy-to-use data collection system that can record over many years.

MONITORING

Saturometer for identifying gas super-saturation

Gas bubble disease is often present undetected and may lead to significant damage to the fish stock. If fish are exposed continuously to super-saturation, they will suffer, no matter which gas contributed most to it. This new handheld saturometer (Protection class IP65, with electric backlight display, identifies gas super-saturation in water reliably to 200 hPa or mbar pressure difference, with an accuracy of 1 hPa. It is powered by 6 x 1.5 Volt Lithium cells, with integrated capacity monitoring on the meter. The maintenance-free sensor is completely sealed (Protection class IP67, Signal output 0-5 V). An optional carbon dioxide and optical oxygen sensor can also be connected to it.

Experiments have shown that continuous exposure to only low levels of super saturation has a significant adverse effect on the fish immune system. A fish stock infected with IPN (Infectious Pancreas Necrosis) suffered losses of 10-15 % after 90 days, whereas fish from the same stock exposed to super saturation showed a mortality rate of 65 %. The gas bubbles themselves cause a lot of direct damage like lesions and embolism. Those alone can be lethal in some cases. But in every case they open the door for secondary infections which cannot be repelled by a weakened immune system. Another typical example is a trout breeder who reported fin-rot and fungus (*Saprolegnia*) among his fish. Measurements showed a gas pressure of 45 hPa, equivalent to 105 % gas saturation.

Waterproof pocket testers

This waterproof (IP67) and pocket-sized testers (153 mm long, x 24 mm diameter, and weighing just 45 g) with digital display are ideal for use in aquaria or aquaculture and other field applications. Different models are available with replaceable electrodes and automatic temperature compensation that measure conductivity or salinity (0-1999 µS EC or 0.0-199.9 ppm TDS), pH-value (0.00-14.00), and temperature (0.0-50.0 °C) or Redox-potential (±1000 mV ORP). Power is from 3 x 1.5 V batteries.

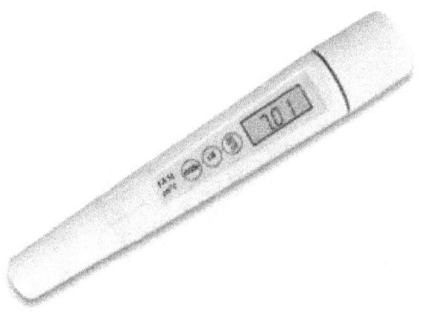

These instruments are calibrated at the factory, but can be manually re-calibrated. Measurements are very accurate (0.1 or 2 % or mV) with high resolution (1.00 to 0.01). The modular design allows easy replacement of electrodes and batteries. A rugged splash proof and floating casing prevents water infiltration.

MONITORING

Photometer for measuring a variety of water quality parameters

To analyze a water sample, simply select the required program by pressing the foil pushbuttons on this microprocessor photolyzer. The display shows the parameter and dimensions exactly (up to 2 decimal places). A signal sounds when the reaction is complete. The results are displayed on the meter and can be stored for later processing. The meter is equipped with an RS-485 adaptor, and can be connected to a printer or PC through the integrated RS-232 adaptor (cable available if needed). It is possible to archive over 500 data sets from up to 16 tanks, each showing date and time.

The meter has a 0-point-memory, which needs no new calibration. This digital photometer works with 4 standard 1.5 V DC batteries, with integrated capacity monitoring on the meter. Alternatively accumulators or a connector for external power supply are available (as accessories). Parameters to be measured include:

Aluminum 0-0.5 mg/l, Ammonia 0-0.5 mg/l, Bromine 0-10 mg/l, Chlorine 0-5 mg/l, Chloride 0-250 mg/l; Copper 0-1 mg/l, Hardness 0-500 mg/l; Iron 0-5 mg/l, Manganese 0-0.03 mg/l, Nitrate 0-100 mg/l, Ozone 0-1 mg/l, Phosphate 0-4 mg/l, pH-value 6-8 and Acid-potential 0-5 mmol, etc.;

Chemical test box for water analysis

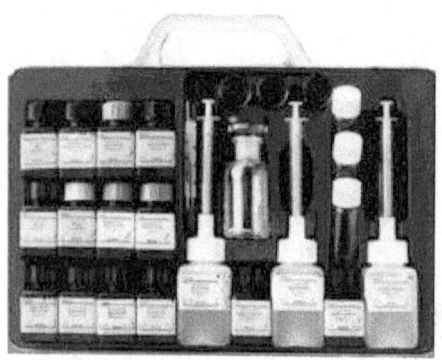

This compact laboratory box-set is equipped with liquid color analytical tests for: Oxygen (Demand + Dissolved), Hardness (Total + Carbonate + Residual), pH, Ammonia, Nitrite, Nitrate, Phosphate, and Temperature.
All tests are also available as single or replacement unit's Additional test sets are available for: Aluminum, Arsenic, Lead, Chlorine, Chloride, Cyanide, Iron, Calcium, Copper, Magnesium, Manganese, Nickel, Sulphate, Sulphite, Zinc, Tin and many more.

MONITORING

Modular multi-parameter measurement and regulation system

When it comes to automatically controlling and adjusting critical water parameters, this system leaves nothing to be desired. Due to its flexible modular concept, the system is the perfect solution for most applications in aquatic environments.

Apart from comprehensive timing functions, it provides measuring modules for the parameters of pH value, redox potential, temperature, conductivity, level (fill level), oxygen and air pressure. Up to eight sensors can be linked to the system in any requested combination. It also allows connecting of up to 16 switching outputs (4x4) for adjustment functions. Measured data storage for 2000 measuring chains, battery backed computer interface for PC-data evaluation with the software, capable of being network-integrated by LAN adapter, flash memory technology for update-function, automatic sensor identification and testing, day/night simulation, lunar phase simulation, high/low tide simulation, flow simulation, timer clock functions, time interval functions, night modus, acoustic, optical/external alarm, etc.

The software is the central and convenient solution for controlling, adjusting and analyzing all measurement and control systems. It allows for analyzing and controlling the data measured by up to 255 industrial systems in one central PC.

Optical oxygen sensor for connection to systems

This new development is the first optical oxygen sensor with SCS interface (Simulation Clark-Sensor), for a perfect integration into existing or running systems. As accessory, a PC-adapter (USB-RS485) is available, with which the sensor can be connected directly to a personal computer, and configures from there (Software for free), where just a net-device is needed and no external meter. The calibration is simple and only needed once as the calibration data can be stored in the sensor. The sensor has no sensitivity towards H_2S or CO_2, but an automatic temperature compensation and malfunction diagnosis. The sensor can be delivered with an IP68 termination plug or with a fixed cable in different lengths. For mounting in a probe holder the sensor can be delivered optionally with a 1" process connector. For fish farming there is a robust version with a biting-protection with nut inserts available.

A display- and regulation system (for up to 8 sensors) is available optionally.

MONITORING

Multi-channel system for control and management

This very versatile instrument can be used as a stand-alone wireless and solar powered multi-channel controller for cages, ponds, tanks and raceways, etc. If there are more than one transmitter units and a base station receiver, then the system automatically configures as a network and sends the information by low powered wireless communication to the base station. This makes for a really low cost simple system and for the first time provides a package ideal for cages, ponds or for any installation that cannot use wires. In the event that you get hit with lightening, then generally only one unit will be affected, and if it cannot be repaired on-site then you simply send the faulty unit back and replace with a working unit. This means that you are in control of your own system and not dependent on engineers.

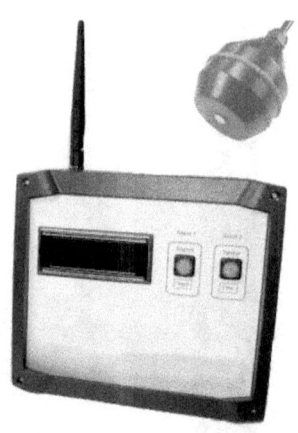

The base station is designed to sit in an office or control room. It communicates with all the transmitters, collects the information and sends onto the PC to data log the information and view the data. The base station has two alarm/control relays so if any one or more of the probes detects a problem, the relays can be activated to trigger your alarm system. The PC does not need to be on for the system to control or alarm. The alarm condition and probe location will be displayed on the base station LCD. One Base station can support up to 50 oxygen transmitters and 100 oxygen and temperature probes.

The transmitter station is a solar powered wireless radio transmitter fitted with LCD display for two oxygen probes, two temperature sensors, and two digital inputs and fitted with two relay control or alarm out-puts. The unit will transmit over a distance of at least 300 m. If longer distances are required a repeater may be fitted to take the distance up to 3 km or more. The transmitter is in a self-contained IP65 enclosure with its own battery and solar cells. The battery alone will power the unit for 3-6 months, but with the solar cells it will not require a battery change for at least 5 years. A 230 VAC/12 VDC uninterruptable power supply, rated at 2 amps with a 5 amp internal back-up battery for the transmitter module which will the battery will keep running for around 30 days, is available as accessory.

The robust and easy to use (60 mm large) oxygen probe is a self-polarizing membrane covered galvanic cell that generates an electrical signal proportional to the oxygen pressure it senses, no matter whether it is in water, air or another medium. It is connected using ordinary cable, which can have any length (up to 1000 m) and does not need an external supply because it makes its own electricity. Cleaning the membrane (with a cloth or tissue) is, apart from an occasional calibration, the only routine maintenance necessary.

BOATDOCKS

BOATDOCKS

Modular floating dock system

25 years' experience in "state of the art" flotation engineering worldwide, together with cutting edge technology and the most advanced production techniques are the base for the new, extremely versatile "Pontoon Building - Block System". By utilizing the friction co-efficient achieved by independently controlled flooding of the modules we have refined this system to provide not only the most stable floating platform available, but also a high load capacity, which can also be combined with a fender-sidebar system.

The individual modules are easy to handle (weight: 6.2 kg) and have a very high load capacity of 360 kg/m² (=75 lbs./sq. ft.). They are made of a high-quality synthetic, UV-stabilized and anti-static material (green or blue), which is extremely resistant to the elements! Individual blocks measure 50 x 50 x 50 cm (1.65 x 1.65 x 1.65 ft.) and the standard freeboard is approx. 40 cm. Blocks are easily assembled (with an assembling wrench) by joining the connecting lugs with bolts that give a secure and stable connection.

These building blocks can be assembled to meet any desired layout or dimensions, and can be used to replace or extend existing pontoon systems. This seawater- and acid-resistant system requires no maintenance or cleaning and is extremely durable. It can be utilized throughout the year or the single modules can be stacked and stored. Unlike closed floats, this system does not blow up in the sun and is resistant to attacks from *Teredinidae*.

Unlike conventional wooden floats, the pontoons are made of a skid-proof material, and will not rot or develop sharp edges, or expose rusty nails etc. The "Maritime Technic - Pontoon System" contributes to water purity and is aesthetically pleasing. It has lugs (ears) along the sides where different accessories can be fitted and anchoring systems attached. It is also possible to attach cage nets on these ears.

Swimming ladders, safety rails, anchor fastening and boat mooring-eyes are the main accessories, with options such as side screws and bars as well as fenders, ropes and anchors. If needed, access ramps or gangways can be produced to the customer's design. For marinas electricity distributors (w/wo illumination) and water supply-systems are available.

BOATDOCKS

Multi-purpose aluminium boats

These boats are made of seawater-resistant aluminum, welded with modern technologies, and need no antifouling. Built-in flotation chambers make them unsinkable. Options and accessories are available.

Type A

Type C

The smaller recreational boats, Type A are of modular construction, and can be disassembled for transport (modules measure 125 x 135 x 55 cm) or lengthened by additional hull sections.
The larger working boats Type C are double-skinned and are verified by Lloyd's Register in Hamburg. They have the EU-conformity certificate (CE-Module-C).

Type	230A	350A	480A	470C	610C	750C
Length cm	230	350	480	470	610	750
Width cm	135	135	135	180	180	180
Height cm	48	48	48	75	75	75
Weight kg	45	60	80	250	300	400
Capacity kg	200	250	350	450	600	800
Parts	2	3	4	1	1	1
Motor HP (max.)	6	10	15	40	60	80

Inflatable work boats

These inflatable boats with thick, durable liner are made to meet the requirements of today's customers. High quality materials and careful production guarantee safety and long life in both fresh and salt water. The large tube diameter (35-56 cm) with 2-6 air chambers and the special design guarantee high stability and safety on the water.
The boats are available in sizes ranging from 200 to 577 cm in length and 134 to 239 cm beam, for loads up to 1000 kg and outboard motors up to 90 HP.

BOATDOCKS

Professional electric motors for boats and barges

An electric motor is practical for boats up to 9 m overall length. In contrast to gasoline-driven engines these motors have very different turning moments, and offer less resistance, which allows the use of relatively smaller power ratings. Experiments have shown that the power rating of an electric motor can be half or less that of a comparable gasoline engine. In all models the casings and fastenings, are made from seawater resistant aluminum. All other metal parts including screws are of stainless steel. All parts including the motor are covered with a special durable weather-resistant plastic coating. They are produced to the highest standards of quality and precision. Each motor is tested to performance.

The shaft models (weight 19-21 kg) are delivered ready to use with cable, fuse and build-in switch, and need only to be connected to the battery. They are available in 2 power ratings (800 or 1400 Watts) with 2 forward- and 2 reverse speeds, and with different shaft lengths.
The bracket-models (weight 15-35 kg) are delivered complete with cable set, fuse and switch, which must be installed together with the pin for attaching steering lines. They are available in 4 power ratings (800, 1400, 2000 and 3000 W) with 4-6 speeds.

Shaft-models

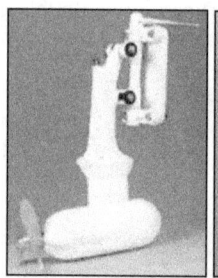
Bracket-models

For all models a special folding propeller is available optionally, which ensures that the motor will not turn if it is not switched on, and reduces flow resistance and noise. As accessories special batteries and speed regulator are available.
On request we supply also petrol or diesel- powered outboard motors.

Polyethylene ropes for permanent connections

These buoyant and shock-absorbing all-round ropes, twisted from polyethylene fibers are UV-, oil- and sea water-resistant. They are available in orange or white, 6, 13 or 20 mm diameter, with working loads of 1 or 3 and 8 tons respectively.
No more splicing: By pushing the rope together a cavity can be created inside the rope, into which the rope end is slid forming stable connection without the traditional interlocking strands. This is perfectly secure, but can easily be opened later.

NETCAGES

NETCAGES

Illuminated net cages for producing fingerlings with natural zooplankton

20 years experience in the field of plankton research and fingerling production with natural plankton, and years of tests under different environmental conditions in marine and fresh water, is the basis for this new cage system.

The method relies on the fish eating naturally produced plankton in lagoons, bays, fjords, lakes and large ponds. The food organisms are attracted to a lamp located in the center of fine meshed submerged nets containing the fish. Predominantly copepods and cladocerans are attracted towards the light source, but other photo sensitive plankton organisms also stream steadily into the cages.

The 2 x 2 x 2 m net cages (volume 8 m^3) are manufactured from highly resistant mono-filament polyester fabric (mesh sizes from 0.4 to 2.0 mm), fitted with a zipper for easy access, and a conduit for the electric cable. The net cage is stretched over a strong metal frame and submerged into the water column. An efficient solar generator can be supplied for remote site operation. Similarly a transformer can be supplied in order to reduce voltage to safer levels for use in water. The system is installed on a floating dock anchored somewhat away from the shore. Suitable sites for such a cage system require waters with a minimum plankton stock of 500-1000 zooplankton organisms per m^3 during the time of operation. Depending on zooplankton concentrations, the fish stocking density (from the fry stage to fingerling size of 5 cm length) can reach 5000 Ind./m^3 and survival rates of more than 90 % have been achieved with various marine and freshwater fish species.

Circular net cages for all conditions

We supply a wide range of single, double and triple ring cage collars, built from black, high density polyethylene (HD-PE 80-100) pipes in sizes from 200 to 450 mm Ø, optionally mount with either dip-galvanized steel or injection molded plastic brackets.

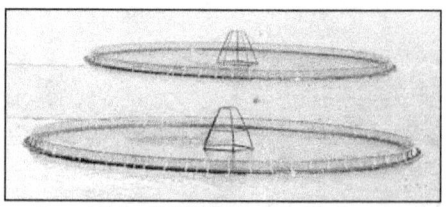

All are built with heavy gauge handrails and uprights (110-160 mm Ø) that resist deformation. These custom-made circular cages range from 10 to 60 m in diameter (32-189 m circumference) and can be built for all types of weather- and sea conditions from sheltered bays to offshore sites. A safety chain inside the outer pipe ring, and dividing plates welded inside the inner pipe ring, are available to enhance safety. A full circumference walkway/decking is available optionally.

NETCAGES

Submersible wave-proof offshore cages

The polygonal framework forms a three-dimensional nonahedron (9-sided structure – axial diameter ~20 m), with service platforms, passage walkways with railings and removable hand-ropes. The framework houses ballast tanks and compressed air cylinders, permanent buoyancy compartments, balancing tanks, and residual buoyancy tanks that allow compensation for changing loads. A rugged, hot-galvanized coating provides a service life of more than 20 years effectively eliminating the need for restoration and repair. The net cage is suspended within this structure.

The automated underwater feed distribution system is intended for storage and measured distribution of feed for autonomous operation over 7-30 days. The system consists of a water tight, cone-shaped cylindrical, 3000 liter bunker that will hold up to 2 tons of feed (on average), and an automatic distributor that will release a predetermined volume of feed. The upper part of the feeder houses solenoid valves, the system controller and the batteries. The system is made of corrosion-proof steel (6-8 mm thickness) and non-ferrous metals. For submerging and surfacing, the cage it is equipped with ballast compartments that are filled by gravity via drain holes equipped with valves fitted with a remotely controlled pneumatic drive. Water from the ballast compartments is displaced by compressed air supplied from the service boat compressor. The process of submerging and surfacing can be supervised by reference to the depth sensor, its display being located aboard the service boat or via an acoustic communication channel from the shore. Flexible pipelines and electric cables run from the signal buoy to the cage supplying compressed air and power. Security is provided by flashing beacons and corner reflectors located on the feed bunker.

The net cage has a volume of 1000 to 2000 m^3 and is suspended within the steel structure by special guy-ropes, running up and down from the service platform. To remove wastes, the net is equipped with a tray device which can be winched to the surface. The cage can be positioned at depths of 4 to 40 meters. Floating at the surface the structure will withstand winds up to force 6 and waves up to 6 meters. Submerged it will withstand force 9 winds and 8-9 meter waves. Operational experience with the system has proved that submergence in the water will effectively protect the structure from corrosion. The system is built and tested according to the "Rules for Building and Classifying Underwater Vehicles, Systems, and Hyperbaric Facilities", and can carry the certificate of the official "Maritime Register of Shipping".

NETCAGES

Underwater lights to increase growth and delay maturity

Underwater lights are used successfully (24 h/day) to delay maturation and increase growth in several fish species (like: Salmon, Trout and Cod etc.) in both sea- and land based farms. This special submersible waterproof (IP 68) lights are made according to IEC and CE standards and are DEMKO and NEMKO approved. The socket of the special lamps is base E40 and the glass lens has a thickness of 5 mm. The light intensity is 70-90 lm/W, the light current is 80-90000 lm and the color temperature 4000-5000 K. They have dimensions (Length x Diameter) of 565 x 130 mm, a weight of 12.5 kg and are delivered with 20-40 m fixed marine cable (3 x 2.5 mm²), and are available in two power ratings (400 or 1000 W, 3.2 or 5.5 A).

Net cleaning system with high pressure

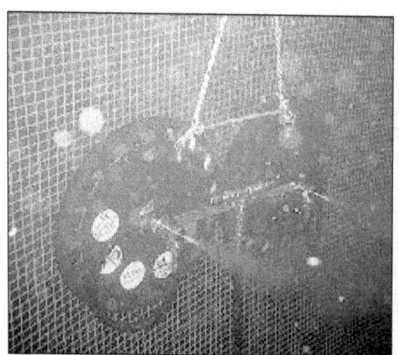

This net-cleaners maintain a flow of oxygenated water through the nets, which keeps them clean with minimal disturbance and no need to remove the fish from the pens. Its construction and design make the washers head working toward the net against its own high pressure water stream and can be used on all types of nets and cages. The net-washer loosen and pulverize, with its high pressure water stream, algae growth, food remains and other fouling, which in turn will drift away by the current.

The single head disk is easily used, the dual head disk is for faster cleaning. The diesel or petrol driven pressure pump with high output drives the head of the net-washer-disk or can be used for pressure washing. It can be operated by one man from the boat or walkway/collar or underwater by a diver.

Traps made from plastic

This traps (with exchangeable components) have a wide inlet opening and an integrated bait basket in the closed end cap. The traps are especially suitable for catching eels, but can also be used for other species. They are delivered complete with sinker weight for submerging, which guarantee that the trap is always positioned with the upper closed side on top, which in turn creates shelter.

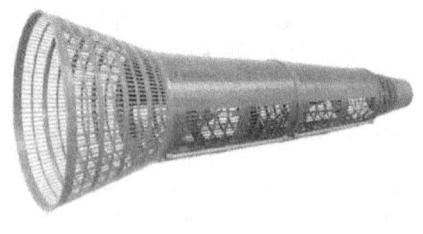

NETCAGES

Precision fabrics for aquaculture and hydrobiology

The fabric program includes precision sieves of different material strengths, and offers a series of mesh sizes ranging from 20 to 2000 micrometer, which will meet all of today's aquaculture and hydrological needs. Our sieve fabrics are offered both by the meter, and as standard- or customized products, such as filter-, dip-, hatching-, cage- and plankton-nets.
In Austria catching zooplankton for feeding fry and fingerlings has been commonly practiced for more than 50 years. Over the past few years several important improvements have been made that offer a better net construction and more effective results. With appropriately designed plankton nets, harvests of several tons per day can be achieved, and with these new nets it is now possible to grade the plankton into 3 different size classes. The result is the correct size of plankton for all development stages: larvae, fry and fingerlings. The expensive and complicated production of prey organisms: algae, rotifers, brine-shrimps and water-fleas etc. is no longer necessary.

Polyethylene shade cloths and protection nets

These shades are made from UV-stable polyethylene material and will protect fish and ponds from leaves, direct sunlight, and bird attacks, all at the same time. The sheets are available in rolls, 1.5 or 3.0 m wide, or can be manufactured to any specified size with extra-strong hem and eyes. Pricing is based on roll wide.

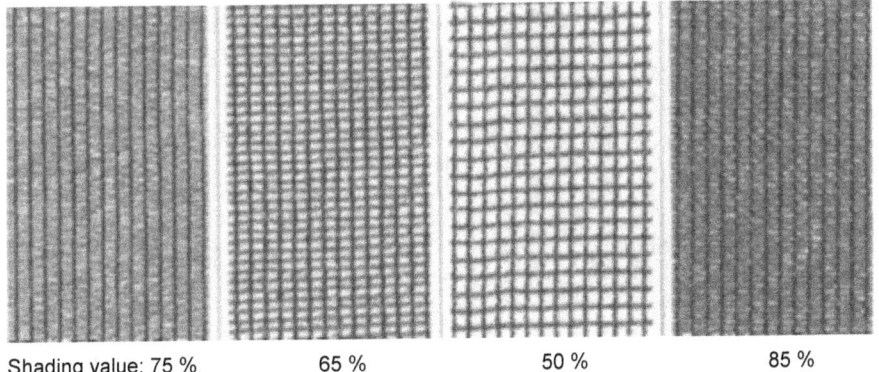

Shading value: 75 %   65 %   50 %   85 %

We supply polyethylene bird protection nets in any length with a width of 10-24 m and a mesh size of 12-30 mm.

NETCAGES

Polymer-coated polyester nets for cages

This special polymer-coated polyester netting is very strong and stable, but also quite flexible. The nets are twice as strong as conventional nylon nets, highly UV-resistant, suitable for us in seawater and need no antifouling. They need less maintenance and allow fewer fish escapes than other nets. Nets have been in use for more than 20 years, and save a lot in operating costs.

The material is available in rolls 3.6-3.8 m wide with a mesh size of 5 or 12 mm, and 5.0 m wide in mesh sizes of 20 and 40 mm. Weighing 450 to 550 g/m^2, the cages need no lead sinkers.

We deliver cages or just the material in rolls with the following specifications*:

Type	45/45-40	50/50-20	40/40-12	40/40-5
Square mesh size (open area)	40 mm	20 mm	12 mm	5 mm
Weight/m2	490 g	550 g	500 g	450 g
Tensile strength (DIN 53857) Warp	45 kN/m	50 kN/m	40 kN/m	42 kN/m
Tensile strength (DIN 53857) Weft	45 kN/m	50 kN/m	40 kN/m	41 kN/m
Expansion force	10 %	12 %	15 %	15 %

*All data are average (+/- 10 %).

Nylon nets for holding and catching

Production of these nets is based on a complete process from selecting the raw material, through the different manufacturing stages: matching, weaving, cutting, assembly and packing. Our aim is to ensure the customer gets a quality product. Woven in nylon, all our nets are sewn with double or triple rows of stitching. Usually knotless netting is used, but on request and depending on use, also knotted netting is available. The finished nets are delivered in any shape or size. Mesh sizes of 4, 6, 8, 10, 12, 15, 18, 20, 25, 30 and 35 mm are available.

Special UV- and antifouling impregnations (green or transparent) are available. These guarantee a longer life for the nets as they do not need to be cleaned as often as those that have no antifouling. On request we can deliver the nets in a range of colors.

NETCAGES

Seine nets available in two designs

Seine net with bag - consisting of two wings with a bag in between, for catching fishes in lakes, rivers and ponds etc.
Seine net without bag - consisting of a well vault single wall, for harvesting fish in ponds, raceways and tanks etc.
Both nets are complete ready to use mounted in strong ropes, with foam floats and lead sinkers.

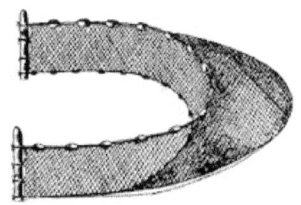

Cast nets available in two techniques

Traditional technique - the lower line is fitted with harvest bags, which retain the fish.
Yugoslavian technique - the lower border is formed into bags by ropes when the net is lifted.
Both nets are delivered ready for use. Circumference is about 7 m, height is approx. 1.7 m, and mesh size is a standard 11 mm

Scoop nets (dip-nets) available in two shapes

D- or O-shaped: strong, corrosion-free scoop-net frames (width: 40-70 cm) with protection and fitting for a handle.
Both nets are complete ready to use mounted with net bag (mesh size 4-20 mm). A hardwood handle can be supplied as an accessory.

Trap nets available in two types

Fish traps - with one inlet funnel, three galvanized iron rings, and two wing nets each 1.5 m long.
Fish bags - with two inlet funnels, five iron or plastic rings, and two wing nets each 2.0 m long or a 3 m long guide net.
Both nets are complete ready to use.
Longer wing or guide nets are can be supplied.

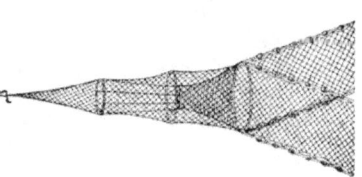

Gill nets available in two versions

Single or triple sheet (trammel) gillnets with a mesh size of 6-150 mm, in various heights and lengths.
Both types are available loose, or mounted with lead- and float lines, ready for use.

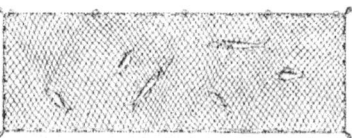

POWERPLANTS

POWERPLANTS

Electric fishing devices with direct- or impulse-current

These safe and powerful electric fishing devices have been used successfully on all continents for over 20 years. Given the different areas, and sometimes problematic conditions of use, a wide range of devices has been developed, which will meet nearly all possible requirements.
All these units are produced according to VDE-regulations and are TÜV-tested. Construction corresponds to security class II. All power-delivering components are double-insulated and guarantee the best possible security for the user. A "Totmann switch" (required in Germany) guarantees additional security, shutting down power to all outlets if a potentially dangerous situation arises. The permanently energized generator and high quality magnets guarantee continuous and reliable power delivery. Each unit will be completely tested, mechanically and electrically, and delivered with a user manual.
Direct current usually creates a better anodic reaction than doe's impulse current, but at extreme conductivities, or for electric barriers, the limits are soon reached with D.C. As result, powerful combined systems (Direct-/Impulse-currents) were developed, where the impulse frequency and voltage can be adjusted as required. The back-pack units are compact and lightweight, with high power output. Combined Direct/Impulse devices are a little larger and heavier than D.C. devices. The weight ranges from 10-135 kg, depending on the model.
We deliver net-, battery- or petrol-driven, back-back- or shore-based units, direct- and/or impulse current, with power outputs of 250-11.000 Watts (5-160 kW/impulse at 25-100 Impulses/sec.), a potential of 300-960 Volts, and for a conductivity of 50-10,000 µS.

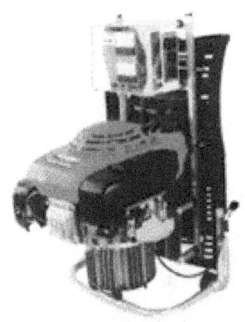  

Back-pack-unit   Stand-unit   Net-unit

The following accessories are available: 30-50 cm Ø stainless steel anodes with RGP-electro handles up to 6 m in length; copper-cathodes with cable and connector; plastic cable drums for up to 100 m rubber coated cable; hand or foot operated emergency-stop-switches; electro-rubber cloths (tested to 1000 Volts) as well as tanks and lights.
Electric fish guidance systems and barriers are available on request.

POWERPLANTS

Hydro-electric generators for low-flow applications

These portable and economic hydroelectric power-plants are suitable for energy generation in the mountains, and in remotely located fish farms.

They are also economic when only small quantities of water are available (from 1 l/sec at 5 m head = pressure 0.5 bar). These compact systems needs no concrete base, and can be placed in the open air. The water supply can be delivered easily by a pipe-line. Water from the power-plant and can be used for other purposes, or discharged back to the stream or river. Water quality is unaffected in any way and the unit could be installed upstream of drinking water- or fish farms sources.

Depending on the water flow, the Pelton-Turbine is driven by water flowing from a regulated (and possible a second, unregulated) nozzle. In turn, the turbine-wheel drives a self-regulated electrical generator. We will adjust the generator size and reduction to the optimal conditions, but these can be easily adjusted later on to meet new conditions. The power produced can be used directly from the regulator and distributor box. To guarantee stable power production when demand and usage vary, electronic power-, water flow- and nozzle regulators are available. The system can be switched off most easily by closing the nozzle(s). All parts are made of durable, anticorrosive materials.

Water cooled diesel generators

These compact, synchronous generators are more powerful than asynchronous generators. Motor, generator and exhaust are water cooled and need no ventilation. The cooling water can be used for heating. They are installed into a housing which guarantees low noise levels (52-54 dB). The units weigh 83-230 kg and are easy to operate with the supplied remote control unit. The diesel fuel consumption is about 0.3 liter per produced kWh. Available in power ratings from 4-16 KVA (3.5 to 14.0 kW). The diesel consumption is about 0.3 liter per produced kWh.

POWERPLANTS

Photovoltaic modules for flexible use

These solar modules (5 mm thick) with 40-cells and seawater resistant outlet, cable and edges have mounting eyes (10 mm Ø) for easy installation and need no back-ventilation. The crystalline electric cells are embedded between elastic liners. The top side guarantees an extreme UV-stability, and is resistant against seawater and bad weather. The back side of the modules consists of a strong stainless steel plate (1 mm thick) to provide stability. The flexible modules can be bent somewhat (up to 3 %) and allow to walk on.

They can be connected in unlimited combinations to reach various power capacities. The following modules are available:

Type/Capacity Wp	18	27	36	54
Size (L x W) cm	48x47	64x47	77x50	91x63
Weight kg	2,5	3,0	4,5	6,0
Power A	1,0	1,5	2,0	3,0
Potential Wh/day	22-72	32-108	42-140	65-216

Wind generators for water aeration or lifting

This system is one of the most modern small size wind generator worldwide. Because of 30-years of experience they are highly economic. The rotor with 6 arms has a diameter of 150 cm and allows a high capacity even at low wind speeds.
Available with a beam of 3 or 6 m in height (6 cm diameter) which is mounted with a rope system (included), it is either supplied with a water pump for lifting of water (suction height max. 7 m, pressure height max. 2 m) or an air pump for aeration of water (incl. 1 m long/deep diffuser).

Generally, the following performances are reached:

Wind speeds m/s	3	4	5	6	7	8
Air pumping capacity l/h	650	870	1080	1300	1520	1740
Water pumping capacity l/h	208	325	375	520	560	600

Alternatively, instead of the pumps, a generator for the production of electric energy is available.

TRANSPORTING

TRANSPORTING

Insulated transport tanks for live fish

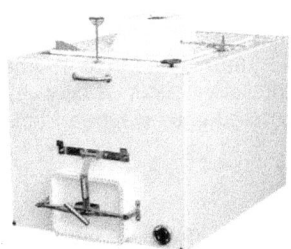

These professional-grade tanks for the long distance transport of live fish are made of strong double-wall, white glass fibre reinforced polyester, insulated with 2 cm thick foam panels, and stainless steel fittings. The tanks have an anti-slip surface on top, which makes it possible to walk and service them safely. For safety, they also have well-fitting splash-proof covers, air and oxygen connections and a handle. The tanks have a 2" outlet and a water-proof hatch for water drainage or exchange, situated on the front.

A large outlet gate with optionally inner closing door allows the tank to be easily emptied of fish and water. As an option these tanks can be supplied with a dividing wall down the middle, and a second outlet gate. As accessories mounting corners, and an outlet chute or funnel, with pipe adapter are available.

Model Type	Volume Liter	Length cm	Width cm	Height cm	Cover cm	Gate cm	Weight kg
TT0800	800	120(137)	90	85(101)	75x60	33x29	115
TT1000	1000	130(147)	100	85(101)	75x60	33x29	125
TT1200	1100	170(187)	105	75(100)	75x60	40x29	150
TT1400	1350	170(189)	105	90(114)	75x60	33x29	165
TT1600	1600	210(227)	105	88(112)	110x75	40x29	175
TT2200	2200	222(241)	105	113(137)	110x75	40x29	205
TT2800	2800	222(241)	105	136(160)	75x60	33x29	215
TT3000	3150	222(241)	105	152(176)	75x60	33x29	235

The numbers in brackets are overall dimension with gate and cover.

Fish transport tanks available in any size

These individually dimensioned tanks for the safe transport of live fish are made of white or green, smooth surfaced, glass fibre reinforced polyester. To use the available space of the transport vehicle more economically, tanks can be manufactured to customer design and dimensions. All tanks are supplied with a well-fitting, removable cover and lid, as well as hand grips. As an option these tanks can be supplied with a slanting inner bottom and splash protecting top and with drainage valve with sieve and outlet gate with door.

As accessories, we can supply and install (12 Volt, DC) circulation pumps or air compressors, pressure reducers with flow meters, or oxygen cartridges etc.

TRANSPORTING

Grading devices for fry and fingerlings

This grading device is especially suitable for hatcheries and small farms. The distances between the grating bars can be adjusted easily by adjusting the grader to form a rhomb (diamond), giving a variable but consistent distance between the bars. The seawater-resistant impregnated wooden frame will float.
Available in two sizes:

Type	Small	Large
Device dimensions cm	36x36x18	45x45x18
Grading distance/thickness mm	4-17	18-30

Grading machines with or without counting system

These graders are made of aluminum and stainless steel and are available for various fresh- and sea- water fishes from Salmoniformes and Perciformes to Siluriformes and Cypriniformes etc. The grading technology is based on 3 diverging V-shaped channels with moving plastic fingers below. These move the fish along the channels until the channel is too wide and the fish fall into to the collecting tanks below. From here they are flushed to the outlets, and optionally by pipes back to the appropriate tanks or holding ponds. Fish are kept wet throughout the process.

The graders are mounted on a 3-wheeled, adjustable trolley. As accessory, automatic fish counting units can be installed at the grader outlets.
The following models are available:

Type	H10	H20	H30	H40	H50
Grades	3	3	3	4	3
Fish sizes	1-50 g	2-100 g	5-500 g	5-1500 g	5-2500 g
Grading capacity	1 t/h	3 t/h	5 t/h	7 t/h	9 t/h
Channel size (LxW)	1600x28 mm	1300x28 mm	2000x42 mm	2500x52 mm	2500x75 mm
Grader outlets Ø	3x125 mm	3x125 mm	6x160 mm	8x200 mm	6x250 mm
Dimensions	3.3x1.0x1.4 m	2.5x0.5x1.2 m	3.5x1.0x1.3 m	4.0x1.0x1.3 m	3.7x1.2x1.5 m
Weight	260 kg	115 kg	190 kg	230 kg	300 kg
Power	0.25 kW	0.25 kW	0.25 kW	0.25 kW	0.55 kW
Water requirement	20 m3/h	15 m3/h	40 m3/h	40 m3/h	50 m3/h

TRANSPORTING

Grading machines with or without fish pump

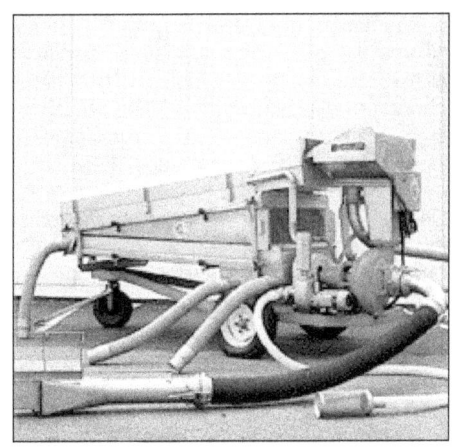

These automatic grading machines for life fishes works with a revolving rollers system and a free adjustable grading process for 3 grads and outlets. Adjustable stainless steel rollers are droved by a 0.5 HP geared motor reducer. A self-priming 3 HP pump with (900 liter/minute) supplies the water, to the rotating rollers and flushes graded fish to the three outlets, to be piped by gravity to the receiving tanks or ponds. It is possible to make 3 contemporaneous selections from 4 to 40 mm thickness. Stainless steel movable discharge outlets that can be used whether on the left or on the right side of the grader.

The standard outfit includes: electrical control panel, flexible suction intake with terminal filter and three flexible connecting pipes for the discharge outlets. All machines are mounted on a hot galvanized steel trolley with three wheels, adjustable supports and shaft. The standard power is 220/380 Volt, 50 Hz (other voltages on request).

The following types are available:

Type ALFA model has dimensions of 195 x 74 x 106 cm and a weight of 195 kg, and is used for fingerlings. The sorting distance can be fine-tuned from 4 to 14 mm only. The grading capacity is 750 kg/hour with medium sized fingerlings.

Type STANDARD model has dimensions of 280 x 110 x 120 cm and a weight of 300 kg, and the fish can be supplied by hand or by an external fish pump. The grading capacity is up to 2,000 kg/hour.

Type GIGANT model has dimensions of 370 x 140 x 130 cm and a weight of 400 kg, and is and the fish can be supplied by hand or by an external fish pump. The grading capacity is up to 5,000 kg/hour.

Type COMBI model has dimensions of 400 x 165 x 180 cm and a weight of 500 kg, and is equipped with a (5.5 HP 220/380 Volt) aluminum fish-pump which draws the fish directly from the pond or raceway, separates them from the water and feeds the grader. This pump embodies a revolutions variator (adjustable by means of a handle), which enables the head and the loading capacity (up to 6,000 kg/hour) to be regulated.

TRANSPORTING

Lifting nets with automatic lock

These lifting nets can be used perfectly to harvest larger fish quantities gently from grow-out- or holding-units. The fish are harvested directly with the lifting net, by crane or hoist, from cages, tanks or other systems where the fish are already kept at high density. For opening the bottom of the net, simply pull the rope of the automatic lock, and the fish are released gently.
The net frame measures 1 m in diameter, and may be rigged with a 50 mm-deep outer net, with or without a tarpaulin liner, and with or without a 10 mm mesh inner net.

Lifting device for live fish

These lifting devices work on the "Archimedes Screw" principle (38 cm Ø) and are suitable for moving up to 6 tons/h of any fish weighing between 3 and 3000 g. They can be used for transferring live fish, from and between ponds, raceways or net cages - to graders, transport tanks or for slaughtering etc. The design offers easy regulation of the screw angle, and they can be easily moved about the farm or hatchery.
They are complete with: screw, frame, wheels, motor (380 V, 0.7 kW), and are available in lengths of 4, 5, 6 or 7 m for lifting heights of 2-4 m.

Vacuum pumps with elevator

With two galvanized- or stainless steel, 200 liter vacuum tanks that alternately suck and discharge, this adjustable fish pump is used for loading transport tanks, and can lift fish up to 6 m. The fish are separated from the water by a grill separator and directed into a container connected to a weight indicator with electronic digital display, that is programmable either by the quantity loaded or the working time. Unit dimensions are 165 x 430 x 340 cm, maximum lift-height is 5.1 m, and weight is 1.5 t. Up to 15 t/h of fish weighing up to 1.5 kg each can be moved.

TRANSPORTING

Vacuum pumps with or without scale

For the lifting of fishes to grading machines and/or the loading of transport tanks etc. this fish vacuum pump is especially suitable. The functional principle is relative easy but effective, which helps to reduce unnecessary breaks and expensive repairs. The fishes (up to 5 kg in body weight and up to 16 t/h) are sucked up to 5 m in height, via an inlet funnel over a suction hose (6 m long) by two vacuum tanks (each 150 l) which work in exchange.

By a 0.6 m² large separator with 8 mm grid distance, the fishes are separated from water and glide so, in the optionally connectable scale or directly into the grading machine etc. The fishes leave the unit by a hose connection (200 mm Ø), which can be adjusted in height from 90-150 cm. The live fish pump has dimensions (L x W x H) of 360 x 115 x 230 cm) and a weight of approx. 650 kg. The motor (380 V/50 Hz, 4.0 kW) is integrated into the housing. The 3 transport wheels and telescopic legs guarantee that the unit stands well and safe.

Impeller pumps with or without separator

This newly developed impeller fish pump is especially suitable for loading live fish of 5-500 g from production units up to 6 m in height to grading machines, holding basins or transport tanks. It works with two electric motors (0.75 and 5.20 kW - 400 V/50 Hz) for fast prime- and main drive (200-750 RPM), and a directly connected maintenance-free gear drive and reaches a fish loading capacity of up to 8-10 t/h. The inlet- and outlet connections have a diameter of 150 mm (6"). The electric box can be controlled directly on the panel or by the supplied remote control unit.

The fish pump made from stainless steel is mounted on a moveable trolley made from aluminum, fitted with 2 inflatable tires (40 cm diameter) and an adjustable stabilizer foot, as well as 2 handles. It has overall dimensions of (L x W x H) 176 x 86 x 106 cm and a weight of 200 kg, delivered complete with 2 camlook units and an intake strainer. Flexible hoses and a separator unit which separate fish from water are available as accessories.

TRANSPORTING

Fish counters for graders or pipes

This newly developed fish counter series made of aluminum and plastic was specially developed to count fast (about 3-5 t/h) and accurately (over 98 %) live fish like salmonids, cyprinids and percids (from 1-5000 g). The counter is simply connected to the outlet pipe of the grader or any other tube. Provided with programmed intelligence, the fish counter performs various controls (auto-test) and informs the user in case of failure. The graphic display and the power supply cable are also equipped with an electrical connector (90-220 VAC/15 VDC). All electronic parts are sealed, easily removable and replaceable by the user. The counters are available in 4 sizes for fresh- or seawater use and supplied with a set of separators and two support legs. A reception tank/funnel with 160/200 mm diameter pipe connection are available as accessories.

Fish scanners for large fish numbers

These fish scanners operate on a patented counting principle that gives both high capacity and accuracy (98-100 %). The fish flow freely through a pipe and pass an advanced camera/measuring system (CCD camera chip) which will register the size and speed of the passing fish. The resulting data are then transmitted to a control unit which can handle up to 4 channels simultaneously. It is not necessary to have the fish pass singly; the counter can separate and count fish that pass simultaneously.

The control unit (dimensions W x H x D 28 x 13 x 37 cm, weight 5 kg) has a powder-painted marine aluminum case, which is sturdy, splash proof, and user-friendly. The power supply is 100-240 VAC 50/60 Hz, 12 or 24VDC. It can power and control up to 4 CSE or 2 CSF registration units at the same time. It has fluorescent control buttons, a backlight display, and connectors for auxiliary equipment like PC, and PLC control etc.
The CSE registration units are used for "dry"-counting (pipe 0-25 % full of water flowing through the unit) and suitable for many applications.
The CSF registration units are used for wet-counting (pipe 100 % full of water) and suitable for special applications.
These registration units are available in different sizes (DN 150-350), weights (14-80 kg) and power consumptions (10-33 Watts), for different fish sizes (1-10,000 g) and capacities (10-100 t/h). All registration units are delivered with 10 meter cable.

TRANSPORTING

Crane scales with swivel hook

These stable and professional quality hanging scales for use with a crane are cast from light but robust aluminum (approx. 260 x 195 x 210 mm), are equipped with a 360 degree swivel security-hook, which is produced for 150 % over load and 500 % breaking strength. The display has large, 30 mm illuminated numbers and an automatic zero point and tare-function. The system will deliver an accurate weight even if the load is swinging. Power is supplied from a built-in battery, delivered with charger. All models can be operated with a remote control. The following types are available as standard:

Type	CS-3000	CS-5000	CS-9000
Capacity kg	3000	5000	10000
Intervals g	1000	2000	5000
Weight kg	14	24	35

Universal scales with flat platform

These new bottom scales with stable base, 4-cell measurement technique and corrosion free platform made from stainless steel (70 mm high) are universally to use. The display (27 x 9 x 4 cm) with 25 mm LCD digits, 100 % tare function and 1.8 m spiral cable can also be mounted on a wall (angle adjustable). It is possible to select continuous display or automatic switch off. The power supply over a rechargeable accumulator with charger allows a mobile use. They have two inserted transport wheels on the side and on the opposite side a hand grip. Optionally a wireless display is available. Following types available as standard:

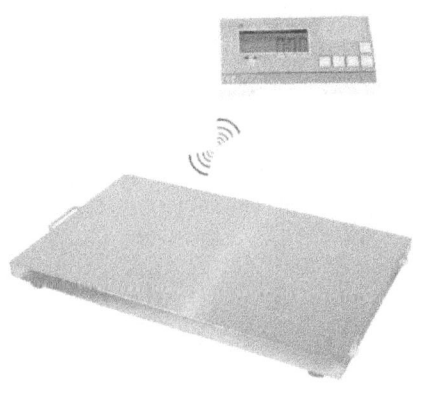

Type	VS-150	VS-300
Capacity	150 kg	300 kg
Intervals	50 g	100 g
Platform	900 x 550 mm	

We deliver also: Precision-, Analyze-, Bottom-, Table-, Desk- or Pocket-scales, for ranges up to 10 t or divisions from 0.01 mg on request.

TRANSPORTING

Motor driven multi-purpose pumps

These compact, self-priming pumps are especially suitable for use in aquaculture, agriculture and industry. They will pump sludge, and even accommodate small stones. Powered by a strong and economic single-cylinder - 4 stroke "Honda" gasoline engine, these centrifugal pumps are reliable and environmentally friendly. A transistor-controlled starter guarantees easy starting, even after prolonged intervals. The WT models are suitable for waste water and will accept stones up to 3 cm diameter. Type WM is especially suitable for seawater and chemicals. All pumps are delivered complete with hose adaptors and ground screens.

Type 10

Type 20

Type 30

The following pumps are available:

Type	WX 10	WX 15	WB 20	WH 20	WM 20	WT 20	WT 30	WT 40
Pumping volume l/min	130	240	600	500	850	650	1300	2300
Pumping height max. m	35	40	32	50	32	26	30	29
Suction height max. m	7.5	8.0	8.0	8.0	7.5	8.0	7.5	7.5
Connections mm/inch	25/1.0	38/1.5	51/2.0	51/2.0	51/2.0	51/2.0	75/3.0	100/4.0
Motor kW/HP	1.1/1.5	1.8/2.5	2.9/4.0	4.0/5.5	4.0/5.5	4.0/5.5	5.9/8.0	8.0/11.0
Tank volume liter	0.6	1.2	2.5	3.6	3.6	3.6	6.0	6.5
Consumption liter/h	0.6	1.1	1.1	1.6	1.6	1.6	2.0	3.3
Weight kg	7	10	21	27	26	38	58	68
Length cm	33	33	47	52	52	62	66	66
Width cm	25	27	35	40	40	46	49	49
Height cm	33	38	36	45	45	47	51	51

Submersible plastic pumps for direct current

These submersible plastic pumps are powered by 12 V DC motors:

Type	T05	T08	T12	T16
Capacity l/h	1800	3000	4500	8500
Connection Ø mm	19	24	29	32
Consumption A	1.3	2.5	3.0	5.0

TRANSPORTING

Large volume propeller pumps

These single stage, axial propeller pumps are suitable for lightly polluted water containing suspended solids up to 50 mg/l at temperatures up to 25 °C. The left- turning motors (380/400/500 V, 50 Hz, 1450 RPM, power factor 0.83-0.87 cos.) have protection against incorrect rotation, and waterproof coils. The motor is protected by a glide seal, the axial and radial bearings are water-lubricated glide bearings. The propeller is made from CuSn10 bronze the radial bearings from bronze and stainless steel, and the bolts and screws from stainless steel.

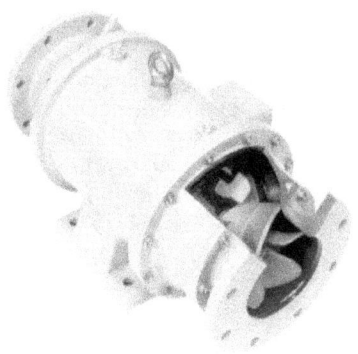

All pumps are made according to DIN EN ISO 9906 class 2, protection art IP68, and are suitable for continuous operation or frequency regulation. They are supplied with 10 m, directly connected waterproof rubber cable. They can be installed at any angle, horizontally or vertically, as well as submerged or free standing.

Type	PO-200	PO-250	PO-300
Capacity m3/h	150-320	0-620	0-1000
Pumping height m	5.5-3.0	9.0-0.0	23.0-2.8
Suction height max. m	1.0	2.0	3.0
Power kW	5.0	9.2	30.0
Weight kg	220	270	590

Multi pumps for many applications

These pumps fitted with plastic vortex propellers can be used submerged (wet) or dry. They have a capacity of max. 20-40 m³/h, a pumping height of max. 6.5-8.5 m, and will accept particles up to 6 mm. Because of their strong ceramic bearings they are suitable for continuous operation. All electric parts are embedded in plastic, and a thermo-switch works as an overload protection. The split-case motor (220-240 V, 50-60 Hz) is low-maintenance and energy efficient.

The large pre-filter caps (front and back) can be removed by hand. The outlet (pressure) connection has inner and outer threads, and is supplied with hose adapters (20/25/32 mm). The pumps in this series are compact: L x W x H of 285 x 170 x 170 to 435 x 240 x 240 mm, and can be used for fresh or salt water. They are delivered complete with 10 m cable and plug.

TRANSPORTING

Submersible pumps for continuous operation

These submersible pumps have special seals and bearings designed for maintenance free continuous operation and have been developed for both mobile and stationary applications in any position. All liquids with a weighing up to 1,100 kg/m^3 can be pumped, max temp: 40 °C, and a pH range of 4-8. The 400 V/50 Hz (2850 RPM) motor is standard, other voltages are available on request. Thermal protection and double case cooling protect the motor from overload (Protection: IP68). The pumps are available with an external or built-in floating switch.

Optional seawater-resistant casings constructed to German Lloyd and Lloyd's Register of Shipping standards are available.

The following types are available:

Type	C-90	C-70/1	C-70/2	B-60/1	B-60/2	A	O
Pumping volume m3/h	0-50	0-24	0-35	0-60	0-80	0-160	0-450
Pumping height m	13-0	15-5	20-0	18-0	25-0	38-0	24-0
Power kW	1.3	1.1	1.5	3.0	4.0	7.5	15.0
Particle Ø mm (max.)	10	5	5	10	5	7	7
Height cm	51	48	48	63	63	74	82
Weight kg	22	24	24	45	45	85	185

Underwater pumps for pipes

These newly designed multi-stage compact submersible pumps are maintenance friendly, and are designed and built for long life. All metal parts: pump housing, cable cover, motor adaptor, wheel, valve, inlet screen and outlet connection, including cable couplings are made from stainless steel (AISI 304), and are completely rust-resistant. Propeller and diffuser are made from polycarbonate, the bearing from polyurethane, and the sealing from a special rubber material.

The maximum outside diameter of these pumps is 99 mm, 4 inches, (3 or 6 inch sizes available on request), so that they fit into a pipe. The outlet connection is 1-2 inch and the length of the pumps is 695-2368 mm and their weight 11-52 kg (depending on types).

We also supply: universal-, circulation-, underwater- and sludge-pumps, in capacities up to 1000 m^3/h, or to over 100 m in height!

PROCESSING

PROCESSING

Electric stunning device for all fishes

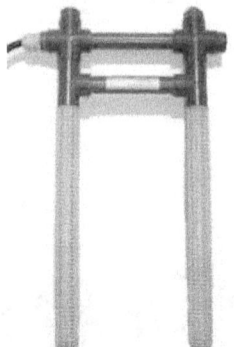

These devices are especially suitable for the fast and easy stunning of fishes (up to 50 kg at a time) prior to slaughter. They contain an electrical transformer (Input 230 V, Output 42 V). All parts are made of corrosion-free materials. The handle with buttons and lights is of plastic; the electrodes are of brass with a honeycomb-like plastic cover. They are available with or without (60 l) plastic container and cover. The fish are placed together with water in a plastic container. To stun the fish, the electrodes are submerged in the water and the button is pressed, a lamp is illuminated when the process is terminated. To reach all fish correctly, the device should be moved around and turned horizontally.

Electric stunning system for much fish

These devices are especially suitable for safe and reliable fish-stunning (up to 200 kg/at a time). They work by electricity (Input 230 V, Output max. 160 V). Three adjustable knobs regulate the pre- and main stunning process and allow the current to be slowly increased, thus and reduce internal damage to the fish. All parts are made of corrosion-free materials. The container and cover are made from white plastic; the electrodes, which are fixed on two sides, are made of stainless steel. The containers are available in 3 different sizes (90, 280 or 610 l volume).

Special knives for the fish industry

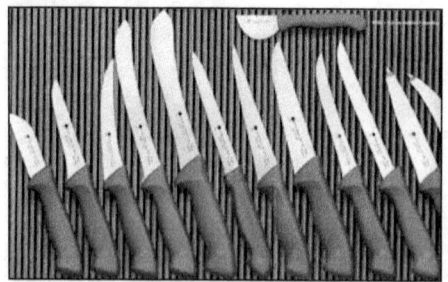

These stainless steel knives are designed for long life and resistance to hard ware. They can be used for many hours without problem to yield high productivity. The blades are hand-finished from high quality steel - a guarantee for maximum cutting precision and easy sharpening. The ergonomic handle design makes them easy and safe to hold.

We prepared special sets (Start, Standard, Profi, and Complete: with up to 30 different knives and blades from 45-315 mm).

PROCESSING

Auxiliary manual fish gutting devices

These gutting devices have been used for more than 20 years and are invaluable for small and medium-sized processing plants. They allow fish between 150 and 3000 g to be gutted and cleaned in a single motion, without letting go of the fish.

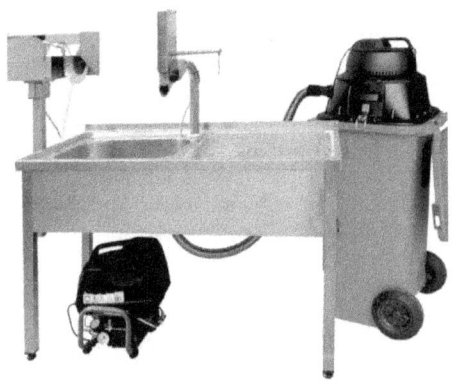

There is no need to turn the fish over, as the sequence of operation starts at the vent. The likelihood of the operator cutting himself is very small. The suction head has two scraper-edges at its lower rim. All the intestines are removed while the device is passing through the fish. By pushing a button, a knife comes down and cuts the esophagus, and the guts are sucked into a container. Optionally, any remaining blood can be removed by the rotating round washing brush (20 cm Ø). A single person should be able to process up to 300-400 fish/h.

Adjustable automatic fish gutting machines

These compact and robust machines are made from stainless steel and plastic. They operate without hydro-pneumatic or electronic elements and are driven by hydraulic pumps, which are integrated with the frame. The machines are nearly maintenance-free requiring only a monthly check of the oil level in a viewing glass, and the exchange of knives or brushes. The lack of complicated and expensive adjustment allow for cost-effective and profitable operation. Incorrect size selection can be avoided with manual size-control (automatic with the Vario model). The speed of the transport chain is as low as possible, which allows the machine to be fed by one hand. A control line on the outlet side allows for an optical control. Bad results (about 5 %) can be removed here. As the machine operates without cutting between the gill covers and pectoral fins, there is no additional waste.

PROCESSING

Automatic de-scaling machines for much fishes

These robust, solidly constructed machines are belt-driven and offer perfect scaling technology. The body is constructed of stainless steel and has an automatic time selector, and a control valve for reduced water consumption. The unit is completely safe to operate, with automatic stop, motor protection, and restart protection).

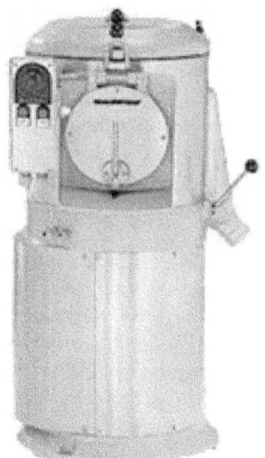

De-scaling takes place at the bottom and sides of the machine and offers the advantages of short (approx. 3-5 minutes per cycle), low-noise de-scaling cycles and reduced water consumption. The water used in the de-scaling process is retained, so that the fish are thoroughly washed and the water provides a soft cushion effect. This ensures that the scaling action is very gentle and the fish are not exposed to heavy bumps or bruised. The auto-stop control automatically stops the machine when the loading lid, top lid, or release flap are opened. Gentle release is ensured because the machine restarts slowly, and the start button must be pressed and held down, which is an added safety feature. The machines are supplied either with a carborundum drum for soft fish or fish with small scales, or with a round-hole drum for clams, hard fish or fish with large scales.

The following types are available as standard:

Type	35 S	16 K	18 K	20 K	25 K
Capacity/cycle kg	4	6	10	15	30
Capacity/hour kg	80	150	280	400	800
Voltage	230	400	400	400	400
Rated power kW	0.18	0.25	0.37	0.55	1.50
Water connection DN	15	15	15	20	25
Depth mm	450	372	530	600	735
Width mm	410	435	555	650	750
Height mm	610	855	930	950	1195

Manual de-scaling device for all fishes

This device is well protected from spray, safe in its operation and securely insulated (tested to 4000 V). The 230 V/50 Hz ICE standard 0.18 kW motor can be used continuously. The machine measures 23 x 24 x 19 cm (L x W x H) and weighs 9 kg. The flexible drive-cable is 160 cm long. The bearings in the hand-held unit with which the fish is scaled, are waterproof. The device is supplied completely with handle, feet and cable.

PROCESSING

Filleting machines for much fishes

This new generation of filleting machines is available in a range of sizes, each suited to different sizes or species of fish (salmons, perch, bass, bream, herring, pollack, mackerel etc.) These permit efficient filleting (approx. 800-1.000 kg fish/h) with minimum loss and high fillet yield (50-78 % from salmonids). The actual yield depends on the freshness and consistency of the fish, its fat content and maturity, as well as the cut of the head (I-, C- or V-cut).

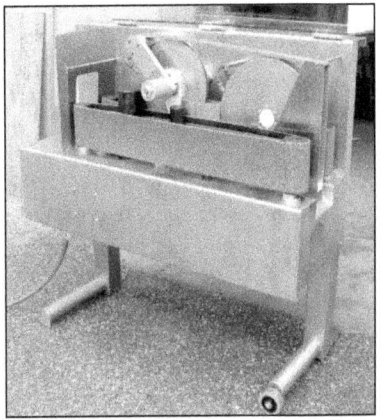

The simple but efficient mechanical action is based on rigorous geometric principles in combination with a system that senses the thickness of each fillet, thus allowing the machine to process fish of differing sizes without changing the settings. The type number represents the maximum thickness of the fish (in mm) measured between anus and dorsum, which the machine can handle. The careful design and sturdy stainless steel construction ensure long life, easy maintenance, and the ability to meet high standards of hygiene. All machines work with 2 or 4 blades (and 1 or 2 electric 400 V/50 Hz motors, others on request).

The de-headed and gutted fishes are loaded by hand into the machine. A 10-12 cm high transport belt carefully transfers (approx. 20 m/s) the fish to the round knife-blades, which run not parallel but axially. The blades cut the fillets and separate the waste bones, thus saving over 70 % of the production time. The filleting machines are available with 2 or 4 blades either for the production of fillets with belly bones (ribs) or de-boned fillets.

The following 2 blade types are available as standard:

Type	Fish Weight kg	Dimensions (LxHxW) cm	Weight kg	Blade Ø mm	Power kW
SF 80	0.1 - 0.5	60 x 54 x 45	65	210	0.55
SF 90	0.2 - 2.0	85 x 120 x 50	92	245	1.10
SF 130	0.3 - 3.0	100 x 125 x 50	110	300	1.50
SF 180	1.5 - 6.0	120 x 135 x 65	170	400	2.20

The following 4 blade types are available as standard:

Type	Fish Weight kg	Dimensions (LxHxW) cm	Weight kg	Blade Ø mm	Power kW
AV 70	0.1 - 0.5	85 x 115 x 50	100	210 + 245	0.8 + 0.12
AV 80	0.4 - 1.5	100 x 114 x 50	145	210 + 300	1.5 + 0.12
AV 100	0.6 - 2.0	120 x 120 x 50	160	245 + 300	1.5 + 0.12
AV 135	0.8 - 3.0	140 x 137 x 50	250	330 + 400	2.2 + 0.25

PROCESSING

Filleting machine for many fishes

This compact machine will fillet all round fish up to a height of 90-135 mm and a weight of 100-1000 g, quickly and accurately, without first gutting the fish. The machine is made of stainless steel and is easily cleaned with water. High output (approx. 100-200 kg/h) and safety provisions allow for fast and efficient operation. The compact design (L x W x H: 118 x 56 x 127 cm) and light weight (approx. 88 kg) enable it to be used almost anywhere. The head is cut off with the rotating round bladed knife, and then the fish is put in the machine head first and belly-up. The fish is advanced by special gripper-belts to the 2 round-blades, and 2 fillets are cut. The cut is adjustable to guarantee low wastes (fillet output approx. 40-50 %). The belly bones have to be removed manually.

Slicing machines for fish fillets

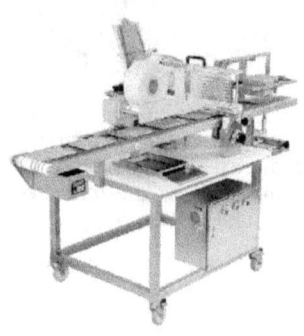

This slicers are based on a proven combination of a sturdy ground machine and a perfect slicing glide. Due to the good quality and the modern technique of the slicers the best slicing results are achieved with yield of approx. 99 %. The plate and glide is from stainless steel, the other parts are made from polished and eloxided aluminum. The blades are made of specially chromed steel and last very long. All this features allow a very economic operation for the processing fillets of various fish species. There are different slicing machines available, (portion or double lane slicer on request).

Type	Manual	Automatic	Compact
Filet temperature °C	-7 to -10	-7 to -10	-4 to +6
Slicing angle degree	10-30	10-30	10-40
Slicing thickness mm	0,5-15,0	1,0-5,0	2,0-10,0
Slicing cuts/min	approx. 60	approx. 60	approx. 75
Power kW	0,25	1,00	1,00
Length cm	62	170	130
Wide cm	60	66	60
Height cm	57	140	130
Weight kg	35	150	120

PROCESSING

Y-bone cutters for fish fillets

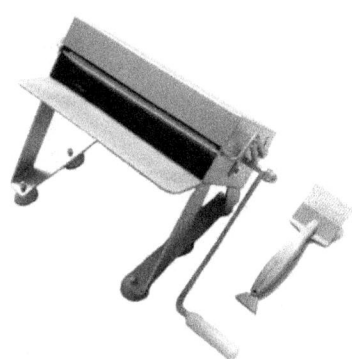

With these compact stainless steel devices (cutting width 9, 15 or 30 cm), you can cut the Y-bones in the fillets from bony fish (like cyprinids, esocids etc.) quickly and easily. By making these cuts at short intervals (every 3-4 mm), the Y-bones are chopped into very small pieces, so that they are no longer detected or dangerous in the fillets, and can be eaten comfortably with the meat. There is also an automatic electric (12 or 24 V DC) Y-bone cutter (cutting width 30 or 40 cm). With an adapter (available as accessory) you can also cut the fillets into chips or strips (and, for example serve it fried or on salad).

There is also a stripe or cube cutter available (i.e. for herring or salmon) which saves about 50 % of working time. The cutting distance can be chosen in steps of 4 mm, i.e. 8, 12, 16, 20 mm (please mention at order).

Pin-bone removers for fish fillets

For removing small and unwanted pin-bones from larger fresh or smoked fillets (150 g or higher from salmon, trout and charr) different pin-boners are available. Machines range from compact handheld devices to automatic belt machines. All machines are built from non-corrosive material. Their unique mechanical picking action gently removes neck- and pinbones, leaving fillets looking their best. They are user- and maintenance friendly and allow fast working. Optionally all belt machines are available as twin models (with 2 heads after each other). The following belt machines are available:

Type	EM-52	EM-102	EM-202
Heads	1	2	4
Capacity kg/h	350-750	750-1500	1500-3000
Belt cm	28	56	112
Wide cm	47	84	115
Length cm	150	180	230

Alternatively, a handheld pin-boner made of stainless steel (weight approx. 1.1 kg) with electric motor and transformer box is available for easy manual operation.

PROCESSING

Skinning machines for fish fillets

These robust fish fillet skinning machines (skinning wide approx. 43 cm, skinning speed approx. 19 m/minute) are manufactured from exclusive, high quality components according to the newest technical and hygienic regulations. They are made of rust-free stainless steel, the transport belts and rolls are of plastic. Additional stability plates inside guarantee a very long life. There are no electronic parts which could become defect. All machines are easy to use and maintenance friendly, which saves operation costs. All models work with an 230/400 V (50/60 Hz) 0.75 kW motor and are equipped with a foot-switch, adjustable knives (the skinning thickness can be adjusted from 0 to 4 mm), rotatable plate, stripper and water spray unit.

Types CF are equipped with a cog-roller (for fish with tough or thick skins) and types CS have a knife- and cleaning-centrifuge (for fish with a soft or thin skin).
The belt-models are suitable for large quantities of fish and guarantee high worker safety during operation (they are switched off automatically if the cover is opened). They have an adjustable pressure unit (opening height max. 90 mm), which adapts automatically to the fillets and is available with different performances, depending on the fish species (soft wheels, hard roller, or soft balloon). With the belt-machines the fillets can be skinned either with a stump blade or with a sharp blade. When skinning with a stump blade (silver skinning) almost no waste is generated, when working with a sharp blade (deep skinning), the skin thickness can be adjusted from 0 to 5 mm.
Optionally, the belt-models can be equipped with specially designed curved blades, which allow the skin and the red-brown muscle segments along the sides of the sensory canal to be removed completely from fresh or smoked fillets. This was one of the most difficult problems to perform by hand. Additionally a removable loading belt, and a forwarding belt are available as accessories.

The following standard types are available:

Type	Table-model		Stand-model		Belt-model	
Type	CF420	CF420	CF460	CS460	CF495	CS495
Width cm	70	70	82	82	70	70
Height cm	42	42	100	100	117	117
Depth cm	50	50	46	46	70	70
Weight kg	70	80	150	160	180	190

PROCESSING

Smoking chambers for closed rooms

These products are fabricated by modern machines and are assembled accurately by hand, inspected and approved. Due to continuing development in close cooperation with the leading professional institutes and associations within the fish- and meat industry, as well as in response to many gastronomic, these are the most highly evolved smokers available; a fact that is confirmed by the best cooks from all over the world, by many professional associations and last but not least by many foreign unions and professional hotel schools. These smokers with their approved technique are in operation worldwide. It is a product for specialty gastronomy as smoking is possible at any time in presence of the customer - also in food-shops, in the grocery stores, at the party-service, and beer-gardens etc.
In the electric pressure chamber griller and smoker you can easily prepare hot or cold smoked specialties ready to serve within a very short time. You can work with it in closed rooms - smoking indoors - always a pleasure. Fish and meat are not cooked directly by firewood or infrared heat, but indirectly by hot air inside the pressure chamber. The tightly-closing door guarantees that only a very small amount of smoke, and nearly no moisture leaks out. Proper gravy will be preserved, which is a great advantage over other smoking processes. Fish are put, belly up, onto the special grates. Flat-fish, fillets, or meat are smoked on the flat grids. The flavor and proper gravy will remain. Cooking is done without the addition of grease or oil, therefore the food is low on calories. Constant temperature and optimal diffusion of smoke (pressure) will ensure the food is properly cooked and has the expected golden-brown color. After the cooking time is completed, smoke will be deposited on the inside of the walls. The darker they are, the better the food taste.

The double wall insulation guarantees very short pre-heating times and very low energy consumption. All electric parts are protected. The smokers as well as the pans and grates are made of stainless steel. Fittings include thermostat, timer, switches, and control lamps, as well as cable and plough.

The smokers are delivered complete with smoking- and drip-pan, 1-2 racks and 1-2 flat grids as well as a manual with recipes.

The following types are available:

Type	HS 24	HS 48
Dimensions (WxHxD)	45x35x45 cm	45x35x85 cm
Racks and grids	30x40 cm	30x80 cm
Capacity max.	24 Fish - 7 kg Meat	48 Fish - 14 kg Meat
Weight	25 kg	45 kg
Connection 110/220 V	1 kW	2 kW

PROCESSING

Smoking cabinets for high product quality

With this smoker you will get a state-of-the-art, high-quality product, finished through CNC-production, and benefit from over 30 years of development and experience. Its most remarkable features are simple operation and compact design. With our special integrated smoke-gas-guidance, your smoked goods will be evenly cooked and aromatized. You will get an optimal use of the smoke, even temperatures inside, and avoid the dripping of condensate. These smokers are made of aluminum-covered steel or stainless steel with interchangeable wood- or gas firing, for hot- and cold smoking of fish and meat, etc. All smokers come complete with drip-pan and bouncer. Grids, hooks, and pipes are available as accessories.
Some of the available types:

Type	RS-20	RS-40	RS-100
Dimensions (WxDxH)	45x36x90 cm	45x36x115 cm	58x44x165 cm
Weight	25 kg	30 kg	80 kg
Capacity approx.	20 Fish or 12 kg Meat	40 Fish or 25 kg Meat	100 Fish or 75 kg Meat
Wood chip use	250 g/Smoking	400 g/Smoking	700 g/Smoking

Smoking ovens for professional use

These electrically heated (380 V/50 Hz) smoking ovens have a capacity of 100-300 fish (equivalent to 70-200 kg meat) for a smoking time of 30-90 minutes. Precise, thermostat-controlled heating (6 kW) produces heat and smoke for slow cooking. A 120 mm smoke pipe connection, and the double-insulated (50 mm thick) walls prevent heat loss and make curing possible in closed rooms. The fish can either be laid on the racks or hung on curing hooks suspended. The ovens are available in 4 different sizes. A built-in pipe connection for an external smoke generator (for cold smoking) is also available. For a continuous operation, trolleys and ramps are available.

PROCESSING

Vacuum packing machines for food

The vacuum chamber and housing of these table models are made from stainless steel and the transparent lid of acrylic glass. Easy and automatic operation is possible through the analog process-time control panel (and in the type "B-35" and onwards, also through the 10-program, digital sensor control). Gas injection system and soft-air ventilations are also available for type "B-35" and onward. The machines have a single sealing-bar configuration (one-time sealing-cutting system available on request) and are available with a double sealing bar configuration from type "B-42" onward. The professional vacuum pump 230 V/50 Hz (vacuum time 10-60 seconds) and back-ventilation guarantee a service-friendly construction.

The following types are available:

Model Type	Inside dimensions WxLxH mm	Outside dimensions WxLxH mm	Sealing bar Length mm	Pump m3/h	Power kW
Mini	280x310x 85	330x450x295	280	4	0.40
Plus	280x310x120	330x450x295	280	8	0.50
Super	350x370x150	450x525x385	350	16	0.55
Jumbo	420x370x180	490x525x430	420	16	0.55
B-35	350x370x150	450x525x385	350	16	0.55
B-42	420x370x180	490x525x430	420	21	1.00
B-50	420x460x180	490x610x445	420	21	1.00

Flake-ice machines with or without storage container

These small, flake-ice makers, made of high quality stainless steel and plastic, have a high production capacity. They are easy to use, and their innovative technology guarantees the stability the ice flakes and very economic operation. All machines are FCKW-free (R 404 a), and have adjustable feet and an automatic water remover.
As an option, a shift key for changing the ice temperature is available which allows the production of ice at either -8 °C or just -2 °C (ideal for fresh fish and seafood).
As accessories, insulated storage containers are available for 70 to 900 kg ice. These keep the ice cool and facilitate easy use, even after longer storage. The two smaller units are also available with integrated storage containers (40 mm insulation) for each 70 kg ice. Larger machines (also without the refrigeration unit), are available on request.

OTHERS

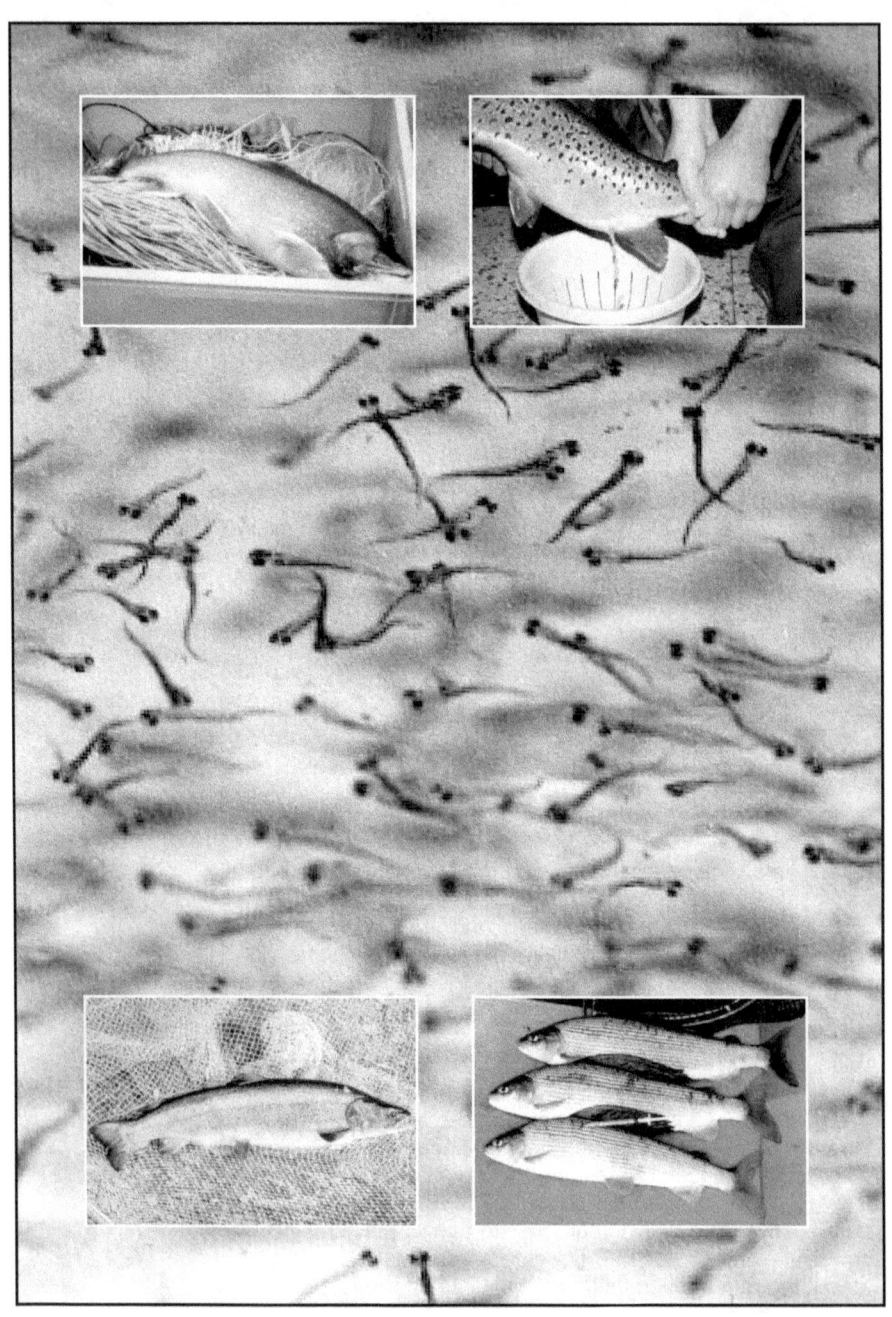

OTHERS

Freshwater fishes with quality

We produce and deliver eggs, larvae, fry and fingerlings of various European freshwater fishes, like:

Whitefishes (*Coregonus lavaretus*):
Various populations and breeds are available. Suitable for stocking in lakes and dams. Systematic stocking of fry in large quantities proves to be most economic. Results achieved in practice have shown that in lakes, profits of up to 100 kg per hectare and year are possible.

Arctic Charr (*Salvelinus alpinus*) and Lake Trout (*Salmo trutta lacustris*):
The brood stock originates from populations from alpine lakes. They are suitable for stocking in lakes, reservoirs and dams, as well as for the production of seafood. The fish are extremely robust and of high stocking value. Tests with tagged fish have shown that fingerlings measuring 3-5 cm are the most economic size for stocking, as adaptation much larger fingerlings to the lakes is more difficult.

European Grayling (*Thymallus thymallus*) and Danube Salmon (*Hucho hucho*):
The brood fish are caught in springtime in several rivers. The fish are bred and fingerlings raised at the nursery farm, initially with live zooplankton and later with industrial dry food. Good results have been achieved with restocking projects.

Pike (*Esox lucius*), Pike-Perch (*Sander lucioperca*) and Burbot (*Lota lota*):
The brood fish are caught in several lakes. Reproduction and raising of fingerlings with live zooplankton takes place at the nursery farm. Stocking of fingerlings at 3-4 cm is most economic and is more as enough in most waters. Excellent results have already been achieved with restocking projects.

Sturgeons (*A. baerii, A. gueldenstaedtii, A. nudiventris, A. ruthenus, H. huso*):
After 30 years of studies and experiments among most known sturgeon species and hybrids, we now offer mainly from our own brood-stock:
Siberian Sturgeon (*Acipenser baerii*): This easy-to-raise sturgeon is grown mainly for meat and caviar production (available from fertilized eggs to selected females).
Russian Sturgeon (*Acipenser gueldenstaedtii*): A beautiful sturgeon (light scutes on dark ground) which is well accepted as an ornamental fish for garden ponds.
Ship Sturgeon (*Acipenser nudiventris*): A rare and interesting sturgeon which builds also freshwater resident populations (in the Ural and Danube River).
Sterlet (*Acipenser ruthenus*): A slow-growing sturgeon which is mainly used for restocking (in the Danube River) and for the aquarium trade (albinos also available).
Beluga (*Huso huso*): The "caviar fish" with the best growth rate: weights of up to 10 kg can be obtained within 2 years and with a water temperature of 20 °C.

We produce over 20 different fish species, including: *Chondrostoma nasus, Barbus barbus, Aspius aspius, Tinca tinca* and others, and can also produce your required species in larger quantities.

OTHERS

Spawn- and foodstuffs for fish production

Carp Pituitary Extracts, are the acetone-dried pituitary glands of carps, which are pulverized and can be used to speed up broodfish maturation (induced maturation) as well as the spawning process itself (induced spawning). Intramuscular injections are highly effective in increasing the incidence of spawning as well as possibly increasing egg-take in certain species. Carp pituitary extract contains a number of pituitary hormones, including gonadotropins, which are effective for the induction of final maturation and ovulation. CPE is administered in distilled water or physiological saline, and injected intramuscularly at a dose of 1-10 milligram/kg fish. Available in vials of 1 gram.

Brine Shrimp Eggs, are the dormancy-eggs (cysts) of *Artemia sp.*, a kind of miniature shrimp (crustacean) which is distributed and sold widely. At present more than 500 Artemia-producing areas are known around the world. *Artemia* are generally used for feeding larval and post-larval stages of fish and shrimps, depending on species, for a period of 10 to 40 days.

The cysts may be conserved for very long periods, and can be hatched as they are needed. The nauplii hatch after 18-30 hours in warm (24-30 °C) salt water (1.5-3.5 % salt), and possess abundant protein and fat. The adults are extremely nutritious, so *Artemia* is an excellent food for a number of fish and crustaceans. They are available as "Premium Quality" in 425 gram (15 oz.) cans, packed at 12 cans per case.

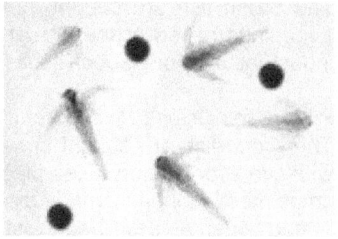

Extruded Dry Food, which has a reddish-brown color and a well-balanced fatty acid profile that promotes good appetite, fast growth, high survival and fish health. The extruded feeds are made from raw material of superior quality and are formulated as a starter- and grower feed especially for closed water or recirculation systems, with minimal phosphate content and low nitrogen excretion.

The particles are also very stable in water and hardly pollute it. The feed has a high energy content, which guarantees very low feed-conversion rates. It can be used from the first day on, or as a replacement feed "artificial plankton" for brine shrimp and zooplankton. It is suitable for the production of a number of fish species such as salmonids, percids, acipenserids, cyprinids and others. On request, larger pellets (6 or 22 mm), pigmented pellets (Astaxanthin 60 mg/kg), or floating pellets (2.5 and 4.5 mm) are available in 10-25 kg bags.

Disinfectants (Chloramin, Virkon, Wofasteril) are also available.

OTHERS

Consulting and planning

Thirty years' of Europe-wide experience and know-how in the field of aquaculture, long-time cooperation with important scientific institutes and researchers, worldwide contacts, and experience-exchanges gained from numerous international congresses and symposia are the base of our respected and successful consulting. Our team can offer cooperation in:

Natural waters –
Consulting and management for fisheries and creation of ecological stocking plans. Physical assistance (also scientific) with restocking-projects for fish, crayfish and pearlmussels, etc. Complete fishing activities with electricity, seine- and gillnets in lakes and rivers. Analyses of plankton and fish populations in ponds and lakes etc. Management of fish stocks in fresh- and seawater. Solutions for parasite problems, slow growing fish stocks or low catch quotas. Suggestions for countering overfishing. Catching of brood-fish, and re-production and rearing of native and difficult-to-rear species and varieties (with zooplankton).

Production systems –
Planning and design for farming and culture systems for freshwater and marine fish species, and shrimps. Calculation and preparation of economic- and feasibility-studies, as well as water- and market analysis, and complete project concepts. Engineering for aquaculture and marine. Management and marketing for fish farms. Education and training of personnel for aquaculture and fisheries.
We have developed a new economic-business plan model, based on a spreadsheet (like MS Works, Excel or Lotus). You can enter all variables like: feeding rates and conversion factors, survival and selling rates, stocking density and tank sizes, prices etc. and it will calculate forecasts of growth and stocks, investment and operational costs and all other requirements.

Sturgeon publications –
The Sturgeons and Paddlefishes (Acipenseriformes) of the World: Biology and Aquaculture. By Martin Hochleithner and Jörn Gessner. Bound, 248 pages. This book is designed to close the information gap between scientists and practitioners. It summarizes current knowledge about all sturgeon and paddlefish species worldwide in a practitioner-oriented working guide.
The Bibliography of Acipenseriformes. By Martin Hochleithner, Jörn Gessner and Sergej Podushka. This is the most important bibliography about Acipenseriformes, and includes nearly all publications from around the world. Over 10000 references are listed alphabetically and cross-indexed to species and headings. Available as electronic E-book (on CD-ROM) or as a printed book (over 500 pages).

OTHERS

www.ingramcontent.com/pod-product-compliance
Lightning Source LLC
Chambersburg PA
CBHW070317230526
45470CB00002B/913